IET SECURITY SERIES 12

Voice Biometrics

IET Book Series in Advances in Biometrics – Call for authors

Book Series Editor: Michael Fairhurst, University of Kent, UK
This Book Series provides the foundation on which a valuable library of reference volumes on the topic of Biometrics is build. *Iris and Periocular Biometric Recognition, Mobile Biometrics* and *User-centric Privacy and Security in Biometrics* are the first published volumes in the Series, with further titles currently being commissioned. Proposals for coherently integrated, multi-author edited contributions are welcome for consideration. Please email your proposal to the Book Series Editor, Professor Michael Fairhurst, at: m.c.fairhurst@kent.ac.uk, or to the IET at: author_support@theiet.org.

Published Titles in This Series:
Published 2018 Drahanský Martin (ed) / Hand-Based Biometrics: Methods and technology
Published 2017 Guo and Wechsler (eds) / Mobile Biometrics
Published 2017 Rathgeb and Bush (eds) / Iris and Periocular Biometric Recognition
Published 2017 Vielhauer / Privacy and Security in Biometrics
Published 2013 Fairhurst / Age Factors in Biometrics Processing

Voice Biometrics
Technology, trust and security

Edited by
Carmen García-Mateo and Gérard Chollet

The Institution of Engineering and Technology

Published by The Institution of Engineering and Technology, London, United Kingdom

The Institution of Engineering and Technology is registered as a Charity in England & Wales (no. 211014) and Scotland (no. SC038698).

The Institution of Engineering and Technology
Michael Faraday House
Six Hills Way, Stevenage
Herts, SG1 2AY, United Kingdom

www.theiet.org

British Library Cataloguing in Publication Data
A catalogue record for this product is available from the British Library

ISBN 978-1-78561-900-7 (hardback)
ISBN 978-1-78561-901-4 (PDF)

Typeset in India by Exeter Premedia Services Private Limited
Printed in the UK by CPI Group (UK) Ltd, Croydon

Contents

List of figures

List of tables

Short biographies of the editors and authors

Editors

Prof. Carmen García-Mateo received the M.Sc. degree in Electrical Engineering ("Ingeniera de Telecomunicación"), in 1987 and the PhD degree 'cum laude' also in Electrical Engineering, in 1993, both from Universidad Politécnica de Madrid (UPM), Spain. She is the director of the Multimedia Technology Group (GTM) at the atlantTIC Research Center. She is currently Full Professor at the "Escuela de Ingeniería de Telecomunicación, Universidade de Vigo".

She has been the principal investigator in more than 25 research projects funded by national or international public institutions and companies. She is the author of 21 original peer-reviewed journal articles, 10 book chapters and more than 75 peer-reviewed papers published in international conference proceedings.

Her research interests are focused on speech technology: speech and speaker recognition, audio segmentation, multibiometrics and sign language recognition.

Dr. Gérard Chollet studied Linguistics, Electrical Engineering and Computer Science at the University of California, Santa Barbara, where he was granted a PhD in Computer Science and Linguistics. In 1983, he joined a newly created CNRS research unit at ENST (now Institut Polytechnique de Paris). He supervised more than forty doctoral thesis. CNRS decided last July 2012 to grant him an emeritus status. He consults for companies such as Intelligent Voice Ltd, Speech Morphing Inc and Zaion. His main publications are available from http://scholar.google.co.uk/citations?user=NakTCiYAAAAJ&hl=en

Chapter 2: Fundamentals of voice biometrics: classical and machine learning approaches

Dr. Alicia Lozano-Diez received the double degree in Computer Science Engineering and Mathematics (2012) and the Master in Research and Innovation in Information and Communications Technologies (2013) from Universidad Autónoma de Madrid (UAM), Spain. Since 2012, she has been with the Audias research group at UAM. During her PhD, she joined the Speech group (Speech@FIT) at Brno University of Technology (BUT, Brno, Czech Republic) for two research internships in 2015 and 2017. In the 2016 summer, she interned at SRI International (STAR Lab, California, USA). Her research focuses on deep neural networks (DNN) for automatic language and speaker recognition. She finished her PhD in 2018 and got an assistant professor position at the UAM, continuing her research at the Audias group. In 2019, she got the H2020 Marie Curie funding for the project "Robust End-To-End SPEAKER recognition based on deep learning and attention models" and joined the Speech@ FIT (BUT) as a post-doc researcher and will resume her assistant professor position at UAM after the project finishes.

Prof. Joaquin Gonzalez-Rodriguez, PhD (1999) in Electrical Engineering from UPM, founded in early 2000s and co-led the ATVS Biometric Recognition Group up to October 2016, leading from 2016 to 2018 the AUDIAS Research Group. Since July 2011 he is Full Professor at UAM. He led ATVS/AUDIAS participations in multiple National Institute of Standards (NIST) Speaker and Language Recognition Evaluations since 2001, and he is since 2000 an invited member of the FSAAWG (Forensic Speech and Audio Analysis Working Group) in ENFSI (European Network of Forensic Science Institutes). In 2008, he received a Google Faculty Research Award and addressed in Brisbane (Australia) a keynote plenary talk in INTERSPEECH 2008. During the academic term 2010–11, he was Visiting Scholar in the Speech Group at ICSI (International Computer Science Institute) in the University of California at Berkeley. His research interests are focused on speech and audio processing, acoustics, forensic science and financial series modelling.

Prof. Doroteo Torre Toledano received the MS degree (1997) and the PhD (2001) in Electrical and Electronic Engineering from UPM, Spain. He is currently Associate Professor and Director of the AUDIAS research group at the UAM. Previously he had been with M.I.T. as Postdoctoral Research Associate in the Spoken Language Systems Group and with the Speech Technology Division of Telefonica R&D. Prof. Toledano has over 25 years of experience in speech processing and over 100 scientific publications. He has participated in 6 EU research projects and in over 40 national projects. He has also participated in over 15 technological competitive evaluations (mainly NIST evaluations) and has organized 4. His current research is focused on audio, speech, speaker and language recognition.

Dr. Daniel Ramos finished his PhD in 2007 in UAM, Spain. Since 2011, he is Associate Professor at the UAM and a staff member of AUDIAS. During his career, he has performed research stays at the Machine Learning Group at the University of Cambridge (4 months in 2019) and the Netherlands Forensic Institute (1 month in 2011). His research interests are focused on probabilistic machine learning, inference and interpretation in forensic science and speech and audio processing. Dr. Ramos is author of multiple publications in national and international journals and conferences. He has also participated in several international competitive evaluations of speaker and language recognition technology, such as several NIST speaker and language recognition evaluations. Dr. Ramos is a regular member of scientific committees in different international conferences, and he is often invited to give talks in conferences and institutions.

Chapter 3: Voice biometrics: attacker's perspective
Priyanka Gupta received her BTech degree in Electronics and Communication Engineering from Sir Padampat Singhania University (SPSU), Udaipur, India, in

2014 and completed her MTech in VLSI from The LNM Institute of Information and Communication Technology (LNMIIT), Jaipur, India, in 2017. During her MTech, she was a member of the VESD research group at LNMIIT, and her research was based on implementation of cryptographic algorithms on field programmable gate arrays. She wrote two papers in IEEE conferences. Since July 2017, she is a doctoral student at DA-IICT, Gandhinagar, India. She is a member of Speech Research Lab @ DA-IICT and working under the supervision of Prof. (Dr.) Hemant A. Patil. Her research interests include voice privacy in voice biometrics and attacker's perspective. She has recently participated in INTERSPEECH 2020 Voice Privacy Challenge.

Hemant A. Patil received PhD degree from the Indian Institute of Technology, Kharagpur, India, in July 2006. Since 2007, he has been a faculty member at DA-IICT Gandhinagar, India, and developed Speech Research Lab recognised as International Speech Communication Association (ISCA) speech labs at DA-IICT. He has published around 250 research publications in national and international conferences/journals/book chapters. His research interests include speech and speaker recognition, analysis of spoofing attacks, TTS, and infant cry analysis. He has coedited four books with Dr. Amy Neustein (EIC, *IJST* Springer) with titles, *Forensic Speaker Recognition* (Springer, 2011), 'Signal and Acoustic Modeling for Speech and Communication Disorders' (DE GRUYTER, 2018), 'Voice Technologies for Speech Reconstruction and Enhancement' (DE GRUYTER, 2020) and 'Acoustic Analysis of Pathologies from Infant to Young Adulthood' (DE GRUYTER, 2020). Dr. Patil has taken a lead role in organizing several ISCA-supported events. Dr. Patil has supervised 5 doctoral and 45 MTech theses (all in speech processing area). Presently, he is supervising 3 doctoral and 5 masters students. Recently, he offered a joint tutorial with Prof. Haizhou Li during Asia-Pacific Signal and Information Processing Association Annual Summit and Conference (APSIPA ASC) 2017 and INTERSPEECH 2018. He offered a joint tutorial with H. Kawahara on the topic, "Voice Conversion: Challenges and Opportunities," during APSIPA ASC 2018, Honolulu, USA. He has been selected as APSIPA Distinguished Lecturer (DL) for 2018–19 and he has 20 APSIPA DLs in four countries, namely, India, Singapore, China and Canada. Recently, he is selected as ISCA DL for 2020–21 and delivered 8 ISCA DLs in India. Recently, he is invited to deliver ISCA DL during overview session of APSIPA ASC 2020, New Zealand, 7–10 December 2020. He will also deliver a keynote address (as ISCA DL) during IALP 2020, Kuala Lumpur, Malaysia, 4–6 December 2020.

Chapter 4: Voice privacy in biometrics: privacy in paralinguistic and extralinguistic tasks for health applications
Francisco Teixeira received the MSc degree in Electrical and Computer Engineering from Instituto Superior Técnico (IST), University of Lisbon, in 2018. He is currently a PhD student at INESC-ID and IST under the supervision of professors Isabel Trancoso, Alberto Abad and Bhiksha Raj. His research is mainly focused on privacy for health-oriented speech processing, an area that combines speech processing, machine learning and cryptography.

Alberto Abad is Assistant Professor at the Department of Computer Science and Engineering of *IST* and researcher and coordinator at the Human Language Technology group of INESC-ID. He received the Telecommunication Engineering and the PhD degrees from the Technical University of Catalonia, Barcelona, Spain, in 2002 and 2007, respectively. During the last 15 years, Alberto Abad has conducted research in the area of spoken language technology, being the author of more than 90 original peer-reviewed articles published in journals and international conference proceedings. His current research interests include robust speech recognition, speaker and language identification, computational acoustic scene analysis, privacy-preserving speech processing and health-care applications.

Prof. Isabel Trancoso is a full professor at IST (University of Lisbon), and President of the Scientific Council of INESC-ID. She received the Licenciado, Mestre, Doutor and Agregado degrees in Electrical and Computer Engineering from IST in 1979, 1984, 1987 and 2002, respectively. Her research covers many different topics in spoken language processing. She was Chair of the ECE Department of IST. She was elected Editor in Chief of the IEEE Transactions on Speech and Audio Processing, President of ISCA, and Vice-President of the ELRA Board. She chaired the INTERSPEECH 2005 conference. She was a member of the IEEE Fellows Committee. She chaired the IEEE James Flanagan Award Committee, the ISCA Distinguished Lecturer Selection Committee, the ISCA Fellow Selection Committee and the Fellow Evaluation Committee of the Signal Processing Society of IEEE. She currently integrates the Editorial Board of the Proceedings of IEEE and the ISCA Advisory Council. She was elevated to IEEE Fellow in 2011 and to ISCA Fellow in 2014.

Prof. Bhiksha Raj received a PhD in Electrical Engineering from Carnegie Mellon University in 2000. From 2001 until 2008 he was with Mitsubishi Electric Research Labs in Cambridge, MA, USA, and has been a professor in the School of Computer Science at Carnegie Mellon University since 2009. Dr. Raj's research areas span speech and signal processing, automatic speech recognition, machine learning, acoustic scene analysis and, most importantly, from the perspective of this chapter, privacy and security in speech processing, an area that he has pioneered since 2008. Dr. Raj is a fellow of the IEEE.

Chapter 5: Voice privacy in biometrics: speaker de-identification
Dr. Paula Lopez-Otero received the MSc and PhD degrees in Electrical Engineering (Ingeniera de Telecomunicación) in 2010 and 2015, respectively, both from Universidade de Vigo. She joined the Information Retrieval Lab (Universidade da Coruña) in 2017 with a postdoctoral grant, and she currently holds a speech scientist position at ELSA Corp.

She is an author of 12 original peer-reviewed journal articles, 4 book chapters and more than 30 peer-reviewed papers published in international conference proceedings. Her research on search on speech has been awarded with several awards in Spanish Albayzin evaluations (2014, 2016 and 2018).

Her research interests mostly focus on speech technology: search on speech, emotional state detection, speaker de-identification and language learning.

Laura Docio-Fernandez received her MSc degree in Electrical Engineering (Ingeniera de Telecomunicación), in 1995 and her PhD degree 'cum laude' also in Electrical Engineering, in 2001, both from the University of Vigo (Vigo, Spain). She is currently Associate Professor in the Department of Signal Theory and Communications of the University of Vigo. She is a member of the GTM at the atlantTIC Research Center.

She has participated in more than 20 research projects funded by national or international public institutions and companies. She is an author of 26 original peer-reviewed journal articles (15 included in *JCR*), 6 book chapters and more than 50 peer-reviewed papers published in international conference proceedings.

Her research interests deal with the application of machine learning, deep learning and signal processing, to real-world tasks such as audio-visual speaker diarisation, characterisation and identification, automatic speech recognition, multimodal biometrics, extraction of metadata from audio-visual contents and sign language recognition.

Prof. Carmen García-Mateo (see above)

Chapter 6: Voice biometrics: performance evaluation
Prof. Jean-François Bonastre is Professor of Computer Science at Avignon University. He received his PhD on automatic speaker recognition in 1994 and his Habilitation à Diriger les Recherches (HDR) in 2000. He served as the vice-president of Avignon University from 2008 to 2015 and was the head of the Avignon University Computer Science Laboratory, LIA, from 2016 to 2020. He joined the Institut Universitaire de France in 2006. He was President of the ISCA from 2011 to 2013 and President of the Association Francophone de la Communication Parlée from 2000 to 2004. He is IEEE Senior Member and was elected member of the IEEE Speech and Language Technical Committee and IEEE Biometrics Council. He is one of the founders of ISCA's Special Interest Group 'Speaker and Language Characterisation' (SPLC). He served in the Scientific Committee of the Montreal Computer Research Center from 2016 to 2020. More information is available here: https://cv.archives-ouvertes.fr/jean-francois-bonastre.

Prof. Anthony Larcher is Professor of Compute Science at Le Mans University. He received his PhD on automatic speaker recognition in 2009 and his HDR in 2018. He is director of the Claude Chappe Computer Science Institute and the secretary of the ISCA's Special Interest Group SPLC from 2018 to 2023. From 2010 to 2014, he spent four years as a scientist for the Institute for Infocomm Research in Singapore. He is the author of more than 60 peer-reviewed scientific publications and supervised 8 PhD students. His research interests are focused on speech technology, text-dependent and -independent speaker recognition, language identification and human-assisted learning for speech technologies.

Chapter 7: Voice biometrics: How the technology is standardized

Dr. Andreas Nautsch is with the Audio Security and Privacy research group at EURECOM. He received the doctorate from Technische Universität Darmstadt in 2019 ('magna cum laude'), where he was with the biometrics group within the German National Research Center for Applied Cybersecurity. He received BSc and MSc degrees from Hochschule Darmstadt (dual studies with atip GmbH) in 2012 and 2014, respectively. He served as an expert delegate to ISO/IEC and as a project editor of the ISO/IEC 19794-13:2018 standard. Andreas is a co-initiator and secretary of the ISCA Special Interest Group on Security and Privacy in Speech Communication.

Prof. Christoph Busch is a member of the Norwegian University of Science and Technology, Norway. He holds a joint appointment with Hochschule Darmstadt, Germany. Further he lectures Biometric Systems at Denmark's DTU since 2007. On behalf of the German BSI, he has been the coordinator for the project series BioIS, BioFace, BioFinger, BioKeyS Pilot-DB, KBEinweg and NFIQ2.0. He was/is partner of the EU projects 3D-Face, FIDELITY, TURBINE, SOTAMD, RESPECT, TReSPsS, iMARS and others. He is also the principal investigator in the German National Research Center for Applied Cybersecurity (ATHENE) and is the co-founder of the European Association for Biometrics.

Christoph co-authored more than 500 technical papers and has been a speaker at international conferences. He is member of the editorial board of the IET journal on Biometrics and of IEEE *TIFS* journal. Furthermore he chairs the TeleTrusT biometrics working group as well as the German standardisation body on Biometrics and is the convenor of WG3 in ISO/IEC JTC1 SC37.

Chapter 8: Voice biometrics: perspective from the industry

Dr. Marcel Kockmann is Chief Technology Officer at LumenVox. He received his Master's degree in Audiovisual Technology from the Technical University of Ilmenau in 2007 and his PhD in Computer Science and Engineering from Brno University of Technology in 2012. He started his career at Siemens Professional Speech Group, moving to SVOX and Nuance. Since 2011, he built up the research team at VoiceTrust, focusing on creating core voice biometric algorithms and building commercial voice biometric and multi-factor solutions. With the LumenVox merger he focuses now on building full-stack conversational AI speech solutions including speech recognition, voice biometrics and natural language understanding. He is the author and the co-author of around 20 conference and journal papers on speaker verification.

Dr. Kevin Farrell is Director of Biometrics and Security Research at Nuance Communications. He received his PhD in Electrical Engineering from Rutgers University in 1993, where his thesis focused on neural network applications within speaker recognition. He has since worked at Rutgers University, SpeakEZ Inc, Dictaphone, T-NETIX, SpeechWorks, Scansoft and Nuance. His career has been focused on the commercialisation of speaker recognition technology including algorithmic research, software development and customer engagement. He has over 30 publications in the field of speech and speaker recognition in addition to 5 book chapters and 15 issued US patents.

Daniele Colibro is Voice Biometrics Senior Researcher in Loquendo/Nuance since 2001. He received an MS degree in Computer Science from the Politecnico of Turin in 2000 and a Master Degree in Telecommunications from L'Universita dell'Aquila in 2004. During the last 20 years, Daniele Colibro has worked in the area of speech recognition, voice biometrics, language identification, speaker segmentation, device recognition and spoofing detection, contributing to several papers and patents.

Claudio Vair received an MS degree in Computer Engineering from Politecnico di Torino, Italy, in 1994 and a Master Degree in Telecommunications in 1996. He worked as a research scientist at CSELT on HMM training and adaptation, speech signal processing and language modelling. For the last 20 years, Claudio has been working in the Voice Biometric field and he is currently Senior Researcher at Loquendo/Nuance Communications. His research interests include speaker and spoken language recognition, speaker diarization, artificial intelligence and machine learning, and he is the author and coauthor of several papers and patents on these topics.

Dr. Anil Alexander is CEO and Co-Founder of Oxford Wave Research Ltd and his areas of expertise are in voice biometrics, forensic audio processing and speaker diarisation. He has many years of experience developing bespoke solutions for law enforcement, military as well as other agencies both in the UK and around the world. He obtained his PhD in 2005 in the area of forensic automatic speaker recognition at the Swiss Federal Institute of Technology, Lausanne (EPFL), working closely with the Faculty of Law and Criminal Sciences, University of Lausanne, Switzerland. He has been particularly successful in bringing practical award-winning applications of state-of-the-art academic research to commercial product development. He is an affiliate member of the NIST (USA) OSAC Speaker Recognition Committee. He previously chaired the research committee of the International Association for Forensic Phonetics and Acoustics (IAFPA).

Dr. Finnian Kelly, Principal Research Scientist, leads R&D developments in speech analysis, speaker recognition and audio processing at OWR. Finnian joined OWR in 2016 as Senior Research Scientist and has successfully led the team in two NIST speaker recognition evaluations. Finnian was previously with the Sigmedia Research Group at Trinity College, Dublin, where he completed his PhD in 2013 and is Research Associate with the Center for Robust Speech Systems at The University of Texas at Dallas. Finnian has published in (and acts as a reviewer for) many top-tier international conferences and journals and has been an invited speaker at research labs in Europe and the US. Finnian is a member of the research committee of the IAFPA, and an affiliate member of the NIST OSAC Speaker Recognition subcommittee.

Chapter 9: Joining forces of voice and facial biometrics: a case study in the scope of NIST SRE'19

Mohamed Amine Hmani is a PhD student at Télécom SudParis, Institut Polytechnique de Paris. In 2016, he received an engineering degree (Diplôme d'Ingénieur) in signals and systems from Tunisia Polytechnic School. He is currently carrying out his PhD thesis under the supervision of Dijana Petrovska-Delacrètaz and Bernadette Dorizzi. His current research is focused on the (re)generation of post-quantum crypto-biometric keys from facial data, which are keys generated from facial data using algorithms thought to be resistant to quantum computing.

Aymen Mtibaa obtained an Electrical Engineering degree in 2016 from the National School of Engineers of Sfax, Tunisia. He is currently a PhD student at Télécom SudParis, Institut Polytechnique de Paris. He conducts his research within the Electronics and Physics department under the supervision of professors Jerome Boudy, Dijana Petrovska-Delacrétaz and Ahmed Ben Hamida. His research is mainly focused on privacy-preserving speaker recognition systems, an area that combines automatic speaker recognition, machine learning, privacy and security in speech processing.

Dr. Dijana Petrovska-Delacrétaz obtained her degree in Physics (1982) and her PhD (1990) from the EPFL in Lausanne. She was working as a consultant at AT&T, as a postdoc at Télécom ParisTech, and as Senior Scientist in the Informatics Department of Fribourg University, Switzerland. Since 2004 she joined Télécom SudParis, Institut Polytechnique de Paris as an associate professor. Her research activities are oriented towards pattern recognition, signal processing and data-driven machine learning methods, which are exploited for different applications such as speech, speaker and language recognition, very low-bit speech coding, biometrics (2D and 3D face, and voice) and crypto-biometrics (including privacy-preserving biometrics). Her publication list is composed of three patents, two publicly available databases (for speaker recognition and biometrics evaluations) and around 100 publications.

Chapter 10: Voice biometrics: future trends and challenges ahead

Douglas Reynolds is a Senior Member of the Technical Staff at MIT Lincoln Laboratory where he provides technical oversight of the projects in speaker and language recognition and speech-content-based information retrieval as well as consulting with many Government agencies applying speech technology. Dr. Reynolds received his PhD from the Georgia Institute of Technology in 1992 with a dissertation on applying Gaussian Mixture Models (GMMs) to automatic speaker recognition. His current research is focused on application of speech technology to real-world scenarios and domain adaptation of speech systems. Dr. Reynolds has over 200 publications, is a Fellow of the IEEE, recipient of the 2017 MIT Lincoln Laboratory Technical Excellence Award, and is a founding member of the Odyssey Speaker Recognition Workshop series

Craig Greenberg
Craig Greenberg is a Mathematician at the National Institute of Standards and Technology (NIST), where he oversees NIST's Speaker Recognition Evaluation series and Language Recognition Evaluation series, and researches the measurement and evaluation of Artificial Intelligence (AI) and other topics in AI and machine learning. Dr. Greenberg received his PhD in 2020 from the University of Massachusetts Amherst with a dissertation on uncertainty and exact and approximate inference in flat and hierarchical clustering, his M.S. degree in Computer Science from University of Massachusetts Amherst in 2016, his M.S. degree in Applied Mathematics from Johns Hopkins University in 2012, his B.A. (Hons.) degree in Logic, Information, & Computation from the University of Pennsylvania in 2007, and his B.M. degree in Percussion Performance from Vanderbilt University in 2003. Among his accolades, Dr. Greenberg has received two official letters of commendation for his contributions to speaker recognition evaluation.

Preface to *Voice Biometrics*

This book is the latest in the 'IET Advances in Biometrics' Book Series. In fact, the history of this Series dates back to the publication by the IET in late 2013 of 'Age factors in biometric processing', which provided the impetus and set the pattern for an on-going Series of books, each of which focuses on a key topic in biometrics. Each individual volume brings together different perspectives and state-of-the-art thinking in its topic area, shedding light on academic research, industrial practice, societal concerns and so on, providing new insights to illuminate and integrate both specific and broader issues of relevance and importance.

The human voice is a natural and obvious candidate for adoption as a biometric modality, since the characteristics of the human voice depend so much on both physiology (shape and structure of the vocal tract, for example) and behaviour (movement of the vocal components in the formation of speech sounds, for example) which determine the sounds emitted by a speaker, but which are generally very characteristic of an individual. It is hardly surprising, therefore, that biometric analysis based on the human voice is a modality which has a long history and which is now well established. Not only that, but voice-based biometric measures offer a variety of positive features such as a high degree of user acceptability, the possibility of action over a distance, no requirement for physical contact with a sensor and so on. But the use of vocal sound production also presents many challenges, of which the susceptibility to background noise or unpredictable variations in basic characteristics as a result of temporary or longer-term physical changes in the state of the individual are obvious examples. It is therefore highly appropriate that we devote one of the volumes in this Series to the human voice and its use in biometrics-based applications.

This book will explore specifically current developments in voice-based biometrics, and will provide a wide-ranging collection of state-of-the-art contributions which survey, evaluate and provide new insights about the range of ways in which available and emerging techniques can now enhance and increase the reliability of security strategies in the diverse range of applications encountered in the modern world. The contributors to this volume are all highly respected experts in the field and are drawn from a variety of backgrounds. As a consequence, the volume as a whole represents an integration of views from across a wide spectrum of stakeholders including, of course, contributors from both academia and industry. We hope that the reader will find this a stimulating and informative approach, and that this book will take its place within the Series as a valuable and important resource which will support the development of important and influential work in this area for some time to come.

Our aim throughout the development of all the titles in the Series has been to inform and guide the biometrics community as we continue to grapple with fundamental technical issues and continue to support the transfer of the best ideas from the research laboratory to practical application. It is hoped that the Series will prove to be an on-going primary reference source for researchers, system users, students and anyone who has an interest in the fascinating world of biometrics where innovation is able to shine a light on topics where new work can promote better understanding and stimulate practical improvements. To achieve real progress in any field requires that we understand where we have come from, where we are now and where we are heading. This is exactly what this book and, indeed, all the volumes in this Series aim to provide.

Michael Fairhurst
Series Editor, Advances in Biometrics Book Series

About the editors

Carmen García-Mateo is a professor in the department of Signal Theory and Communications of the University of Vigo (Spain) and the director of the Multimedia Research Group (GTM). Her research interests include Speech Technology, Audio Segmentation and Biometrics. She was the recipient of the '2014 Xunta de Galicia Josefa Wonenburger Award' for her outstanding career in the fields of science and technology. She received her PhD Degree from the Technical University of Madrid, Spain.

Gérard Chollet is VP of Research at Intelligent Voice, UK. His main research interests include Phonetics, Automatic Audio-Visual Speech Processing, Spoken Dialog Systems, Multimedia, Pattern Recognition, Biometrics, Privacy-Preserving Digital Signal Processing, Speech Pathology and Speech Training Aids. In 1983, he joined a newly created CNRS research unit at ENST (Telecom-ParisTech within the Institut Mines-Telecom). In 1992 he was asked to participate in the development of IDIAP, a new research laboratory of the Fondation Dalle Molle in Martigny, Switzerland. In July 2012 the CNRS granted him an emeritus status. He holds a PhD Degree in Computer Science and Linguistics from the University of California, Santa Barbara, USA.

Chapter 1

Introduction

Carmen García-Mateo[1] and Gérard Chollet[2]

Voice is the most natural way that humans have to communicate with each other. As a signal, the voice is quite complex, conveying multiple types of information, including, of course, information about the identity of the speaker. Discovering the identity of the speaker is the ultimate goal of the science of voice biometrics, which is also the goal of this book.

The aim of the book is about presenting the reader, whether a student, an engineer, an entrepreneur, a person interested or working in biometrics, the state of the art in voice biometrics research and technology. Nevertheless, we will not only talk about research and technology. Currently, biometrics is a well-established term used not only in the academic environment but also by the general public. Biometric systems, including voice biometric ones, are already implemented in applications of massive use. We are talking about mature technologies that allow their integration in products and solutions because they meet three key requirements necessary for the success in the deployment: highly trustable regarding privacy protection issues, easy to use (ergonomics issues are key factors in their design), and always available (readability). In the book we will also deal with these and other aspects of implementation and deployment (for instance, interoperability and scalability) to which we must pay as much attention as to the performance of the biometric recognition algorithm itself.

The book is structured into 10 chapters; each of the chapters has been written by a team of international experts on the covered topic. Next, the content of each chapter is briefly presented.

Chapter 2 – Fundamentals of voice biometrics: classical and machine learning approaches

In this chapter, written by Alicia Lozano-Diez, Joaquin Gonzalez-Rodriguez, Daniel Ramos and Doroteo T. Toledano, the main state-of-the-art research approaches to

[1]Universidade de Vigo- AtlanTTic Research Center
[2]Intelligent Voice Ltd, CNRS-SAMOVAR, Institut Mines-Telecom, Evry, France

speaker recognition are described. Techniques go from the first successful Gaussian mixture model–universal background model (GMM–UBM) (late 1990s) and the well-established i-vectors (since mid 2000s), until the nowadays state-of-the-art deep learning approaches. Therefore, the focus is set up on the use of deep neural networks (DNNs) and different architectures as feature extractors, as well as their use to replace other modules in traditional systems such as the computation of posterior probabilities or sufficient statistic estimation (instead of the UBM), ending with the most recent trend to develop end-to-end systems based on deep learning techniques. This way, the evolution of the automatic systems for speaker recognition is reviewed.

Among all the voice biometric applications, automatic speaker verification (ASV) deals with verifying the person's claimed identity with the help of machines. ASV finds its applications in telephone-based banking transactions, access of restricted areas/buildings, etc. An important question that must be answered while designing an ASV system is how resistant an ASV system against spoofing attacks, such as identical twins, professional mimics, speech synthesis, voice conversion, and, more recently, replay. Privacy preservation is of utmost importance. The widespread use of cloud computing applications has created a society-wide debate on how user privacy is handled by online service providers. Regulations such as the European Union's General Data Protection Regulation (GDPR) have put forward restrictions on how such services are allowed to handle user data. The field of privacy-preserving machine learning is a response to this issue that aims to develop secure classifiers for remote prediction, where both the client's data and the server's model are kept private. This is particularly relevant in the case of speech. Therefore, the next three chapters deal with the important topic of privacy preservation in voice biometrics applications, namely in ASV applications. Each contribution covers a specific aspect of this crucial issue:

- Chapter 3: Attackers' perspective by Priyanka Gupta and Hemant A. Patil.
- Chapter 4: Privacy in paralinguistic and extralinguistic tasks for health applications by Francisco Teixeira, Alberto Abad, Isabel Trancoso, and Bhiksha Raj.
- Chapter 5: Speaker de-identification by Paula Lopez-Otero, Laura Docio-Fernandez, and Carmen García-Mateo.

A more detailed description of each chapter goes next.

Chapter 3 – Voice biometrics: attacker's perspective

Boosting the resiliency of ASV systems by considering the attackers' perspectives has become definitively crucial. In particular, there have been numerous possibilities to attack ASV systems and, hence, their security can be boosted if the attackers' perspectives are taken into account beforehand, uncovering possible vulnerabilities and loopholes. Thus, this chapter intends to provide insights and understanding into the attackers' possible perspectives, potentially helping in the identification of hidden ASV systems' weaknesses. Details on different attacks based on the extent of attackers' accessibility to the ASV systems are presented. Apart from the attacks, the threats due to unprotected speech corpora are also discussed. Unprotected speech corpora and ASV systems enable to search for information about a speaker on the Internet and, in this context, privacy-preserving techniques can prevent attackers from getting enrolled speakers' information. Consequently, this chapter additionally discusses various technological challenges occasionally faced by attackers, allowing for their positive exploration to come up with better defense mechanisms.

Chapter 4 – Voice biometrics: privacy in paralinguistic and extralinguistic tasks for health applications

Voice privacy preservation concerns not only the linguistic contents but also the paralinguistic and extralinguistic information that may be extracted from the speech signal. This chapter presents a brief overview of the current state of the art in paralinguistic and extralinguistic tasks for a major application area in terms of privacy concerns – health, along with an introduction to cryptographic methods commonly used in privacy-preserving machine learning. These will lay the groundwork for the review of the state of the art of privacy in paralinguistic and extralinguistic tasks for health applications. With this chapter we hope to raise awareness to the problem of preserving privacy in this type of tasks and provide an initial background for those who aim to contribute to this topic.

Chapter 5 – Voice privacy in biometrics: speaker de-identification

Speech de-identification is a process by which a data custodian alters or removes individual's identifying information from a speech dataset, making it almost impossible for dataset users to determine the identity of the subject from which the data were extracted while allowing for data re-use and share. This chapter reviews the most common techniques used for speaker de-identification for two broad fields of application: privacy-preserving voice-driven transactions and privacy protection of sensitive information. A discussion about how to evaluate performance is also provided. Finally, a comparison of techniques is performed in an experimental framework.

Chapter 6 – Performance evaluation of voice biometrics solutions

This chapter by Jean-François Bonastre and Anthony Larcher is dedicated to performance evaluation of voice biometrics solutions. The specific aspects of voice biometrics, compared to other speech-based technologies, are presented, as well as their consequences in terms of performance evaluation. The main existing evaluation protocols and metrics are then presented and discussed, with a focus on the speaker verification part of a voice biometrics systems. Other aspects such as calibration, diarization, forensic, or privacy are also introduced. Finally, some limits of performance evaluation and some guidelines are proposed.

Chapter 7 – Voice biometrics: how the technology is standardized

There are a number of real-life applications that can benefit from robust and mature speaker recognition algorithms. In this chapter three of these arenas are presented:

- Forensic Speaker Recognition, by Anil Alexander and Finnian Kelly from Oxford Wave Research Ltd
- Automated Password Reset: An example of a commercial application using voice biometrics, by Marcel Kockmann from LumenVox
- Testing of Commercial Voice Biometric Systems, by Kevin Farrell, Daniele Colibro, and Claudio Vair from Nuance Communications.

Chapter 8 – Voice biometrics: perspective from the industry

This chapter by Andreas Nautsch and Christoph Busch reports on ISO/IEC projects relevant to voice biometrics (2382-37, 19794-13, 19795-1, 24745, 30106, and 30107) and on the research communities' efforts, namely, de facto best practices, international technology evaluations hosted by NIST, and the ASVspoof consortium, that are co-aligned with the community's interests during ISCA Odyssey and Interspeech Conferences, and the voice biometry initiative. Harmonization activities concern vocabulary, data interchange formats, performance testing and reporting (operational to technology and laboratory validation), the data protection of biometric information, the detection of presentation attacks and their assessment, as well as interoperable interfaces in distributed systems.

Chapter 9 – Joining forces of voice and facial biometrics: a case study in the scope of NIST SRE'19

While the other chapters of this book are devoted to voice biometrics, in this chapter (written by Aymen Mtibaa, Mohamed Amine Hmani, and Dijana Petrovska Delacretaz) an example of joining forces of voice biometrics with other modalities is presented. The focus of this chapter is the combination (fusion) of voice and facial biometrics, also called audio-visual biometrics. It is well known that multi-biometric systems have the advantage of improving the accuracy over single systems, providing increased security and making spoofing attacks more difficult. Regarding voice biometrics, combining voice and facial biometrics provides specific advantages that can be exploited in various manners.

Chapter 10 – Voice biometrics: future trends and challenges ahead

Finally, in the last chapter, authors (Douglas Reynolds and Craig S. Greenberg) present their assessment of future trends in a number of key areas. All of them related to the fact that a person's voice is used more to control real-world actions and access private information. So, the role of voice biometrics will play an increasingly important role in protecting sensitive access and actions and providing personalization for services and devices.

Chapter 2

Fundamentals of voice biometrics: classical and machine learning approaches

Alicia Lozano-Diez[1], Joaquin Gonzalez-Rodriguez[1], Daniel Ramos[1], and Doroteo T. Toledano[1]

In this chapter, the main state-of-the-art research approaches to speaker recognition will be described. Techniques go from the first successful Gaussian mixture model (GMM)–universal background model (UBM) (late 1990s) and the well-established i-vectors (since mid-2000s), until the recent introduction of deep learning approaches, which have revolutionized this field and many others such as computer vision or speech processing in general. We will focus on the use of deep neural networks (DNNs) and different architectures as feature extractors, as well as their use to replace other modules in traditional systems such as the computation of posterior probabilities or sufficient statistic estimation (instead of the UBM), ending with the most recent trend to develop end-to-end systems based on deep learning techniques.

This way, the evolution of the automatic systems for speaker recognition will be reviewed. We will highlight the difficulties intrinsic to the task of disentangling the speaker information from the rest of nuisance variability contained in the speech signal, and how automatic systems have been designed to deal with it. We will also present different approaches to both text-dependent and text-independent speaker recognition and the importance of obtaining calibrated outputs for the systems.

2.1 Introduction to speaker recognition systems

The way a person produces speech is a combination of several factors, some of them related to that person's physiognomy, and in general, related to the person's identity. This way, the voice is considered a biometric trait since from the speech

[1]Universidad Autonoma de Madrid, Madrid, Spain

signal we can extract information that is able to identify the person speaking in a given recording.

Humans are able to recognize people from their voices, especially when they are well known by the person trying to identify the speaker. However, automatic speaker recognition (or identification) systems provide a tool to process large amounts of speech data with a reasonable speed of processing the information.

This fact is especially helpful nowadays, when there is an exponential growth of multimedia data that needs or wants to be managed in an efficient way. This data involves a wide range of speaking conditions and comes from very different audio sources such as call centers, mobile phones, broadcast data, and videos from individuals. Automatic systems are able to disentangle the target information from the rest of nuisance contained in the speech signal, most of the time orders of magnitude faster than real time on huge amounts of conversations and it is reliable enough (although not error free).

Furthermore, they give the opportunity to repeat the process and obtain again the results of a given experiment in an objective way, in contrast to what happens if this is done by humans. They are usually based on well-known signal processing techniques, combined with pattern recognition algorithms and avoiding subjective components in the output decision process. They are usually evaluated on large labeled datasets such as those provided regularly by the US NIST (National Institute of Standards and Technology) for the speaker recognition evaluations (SREs), which is a series of well-known technology evaluations that established a common framework for researchers and have notably driven the field trends.

Nevertheless, the task of speaker recognition presents several challenges, mainly due to the large number of sources of variability, both intrinsic and extrinsic to the speaker itself. For instance, the devices used to record and transmit the voice (microphones, telephones, etc.), the environment conditions (noise, reverberation, etc.), other extrinsic factors, and also the health conditions of the speaker, his/her emotional state or the aging, to name a few intrinsic factors. All these conditions might change from the data used to train the automatic speaker recognition system with respect to the data used for its evaluation, degrading the performance of the system and its reliability.

Moreover, this variability information, as well as the speaker identity information, is spread across different levels of the speech signal. This fact has been exploited by automatic systems based on different features from high (prosody, voice quality) to low (short-term *cepstral* features) levels of analysis of the signal.

In order to tackle all the difficulties of the speaker recognition task, during the years, automatic systems have adjusted and improved the techniques benefiting from approaches emerging in different areas, especially signal processing and machine learning. From the successful GMM-UBM models in the late 1990s, to the well-established *i-vectors* in the mid-2000s, up to the most recent approaches based on deep learning algorithms, this chapter will review the progress of this type of automatic systems over the last decades. Although we will focus on the advances in text-independent speaker recognition (and the challenges provided by NIST evaluations), we will also include an overview on the text-dependent variant and some insights about the calibration of the output scores of speaker recognition systems.

2.2   Metrics for system performance evaluation

As we have mentioned before, one of the main advantages of automatic speaker recognition systems is that we can repeat the same test several times, being able to reproduce the same results. Thus, we can assess their performance in an objective way. Moreover, we are able to evaluate different conditions and datasets, from which we know the ground truth speaker identities. This way, we can compare their performances, providing metrics that let us analyze differences among various systems.

In a typical speaker verification task, we will compare two utterances or recordings, and this comparison is known as a *trial*. The output of the system will be a score with the property of the higher the value, the more support to the hypothesis of both recordings being spoken by the same speaker. These scores are then transformed to hard decisions by comparing them with a threshold. In order to evaluate our system's performance, we will compare the decisions provided by our system with the ground truth labels, which will indicate whether those trials are *target* (same speaker) or *non-target* (different speakers).

One of the methods to analyze the discrimination power of our system is to represent the distribution of both target and non-target scores as, for instance, histograms. This could be a first approach to select a threshold. However, target and non-target score distributions usually overlap, and thus setting a proper task-independent threshold is not easy (if even possible), and issues regarding this decision and *calibration* are presented in Section 2.5.

To overcome this evaluation of hard decisions that depend on the chosen threshold, typically system's performance is tested using metrics such as the receiver operating characteristic (ROC) curve, its variant called the detection error trade-off (DET) curve, the summary value of equal error rate (EER) or a detection cost value defined by an objective cost function. We will see each of them in the following sections.

2.2.1   ROC, DET and EER

These three metrics measure the discrimination ability of a system and are based on the same underlying idea: a moving threshold that is compared to the scores to provide hard decisions. For a given value of the threshold, the decision for a trial will be acceptance if the score is higher than the threshold, and rejection otherwise. This would cause two types of errors: false alarm or false acceptance if the decision was acceptance but the ground truth indicates it is a non-target trial; and false rejection or miss detection if the decision was rejection but the trial is target according to the ground truth labels.

This way, as the threshold sweeps, the false alarm and false rejection errors will change. The EER is the point where the false acceptance and false rejection rates are equal (cross each other). This value is typically used to summarize the system performance in a single value. Since it is a measure of the error, the better the system performance, the lower the EER is, indicating a better separation between score distributions.

However, the false alarm and false rejection rate curves can have very different shapes that provide very similar EER values, and the shape of the curves itself is informative of the system performance. The whole shape can be seen in the ROC or the DET curves. They represent false acceptance rate versus false rejection rate for a moving threshold, with a modification of the axis in the case of the DET curve that provides a better resolution for values closer to the origin of coordinates. The DET curves are straight lines for score distributions that are Gaussian. In both cases, the closer to the origin of coordinates these curves are, the better the system performance.

These curves provide a view of the behavior of different operating points for both of the errors (false acceptance and false rejection). This way, we could select the best operating point (and the best threshold) depending on our application, since we might prefer to minimize one of them while compromising the other.

2.2.2   Detection cost function

Instead of evaluating the discrimination power of a system, it is also convenient to test how good the decisions of the system are. Depending on the application, the cost of a false acceptance or a false rejection might be different. And also, the *a priori* probability of facing a target or a non-target trial varies from one application to another. For example, a banking application that grants access to an account will need to be set to minimize the false acceptance rather than the false rejection. If that account is known to be target for attackers, the *a priori* probability of a non-target can be higher than the target. Usually, when we deal with big databases of speakers, the *a priori* probability of a non-target is way higher than the target: we can create much more non-target trials than target trials since the latter require several recordings from the same speaker.

Considering this, a detection cost function can be defined, as it has been done by NIST in the evaluations mentioned in Section 2.1. This cost can be computed as

$$C_{\text{DET}} = C_{\text{FR}} \times P_{\text{target}} \times P_{\text{FR|target}} + C_{\text{FA}} \times P_{\text{non-target}} \times P_{\text{FA|non-target}} \tag{2.1}$$

where C_{FR}, C_{FA} and P_{target} ($P_{\text{non-target}} = 1 - P_{\text{target}}$) are the costs associated with a false rejection and a false acceptance and the *a priori* probability of a target trial, which are defined depending on the application; and the conditional probabilities $P_{\text{FR|target}}$ and $P_{\text{FA|non-target}}$ are system-dependent.

This C_{DET} depends on the chosen threshold. The optimal threshold for the discrimination can be computed and thus, the cost associated to this optimal threshold is usually referred to as $\min C_{\text{DET}}$. When the difference between C_{DET} and $\min C_{\text{DET}}$ is small, the system is well calibrated (see Section 2.5).

2.3 Text-independent speaker recognition

Depending on whether the content of the recording is taken into account or not, automatic speaker recognition systems can be classified into *text-dependent* or *text-independent* systems, respectively. In this section we will focus on the latter (see Section 2.4 for the first type).

In order to develop a text-independent speaker recognition system, we need to use techniques that leverage information about the sounds and the way they are produced by a given speaker, regardless of the language and the message contained in it. This way, these systems focus on extracting information from the speaker that allows to compare different recordings with different messages. To accomplish this challenging task in a reliable way, this type of systems require longer utterances (30 seconds of actual speech in the recording showed good performance in the NIST SRE evaluations) and low mismatch in terms of acoustic and speaking conditions (way of speaking, channel, noise, etc.) between the two recordings that are being compared. The mismatch between recordings lead to degradation in performance [1].

In this section we will describe the techniques that have been developed for the task of text-independent speaker recognition, from the GMM-UBM approach that dominated the 1990s [2] to the joint factor analysis and total variability subspace-based models that conquer the technology from the early 2000s [3]. Finally, the more recent approaches based on DNNs are presented in Section 2.3.2.

2.3.1 Classical acoustic approaches: GMM-UBM, i-vector and PLDA

Classical acoustic speaker recognition systems are composed of a first stage where the speech signal is transformed to a sequence of feature vectors. These features represent the information contained in the signal in a frame-wise basis, ideally focusing on characteristics of the speaker identity. Frequently, these systems use acoustic features based on Mel-filter banks, such as Mel-frequency cepstral coefficients (MFCCs), or perceptual linear predictive (PLP) features, to name a few.

These sequences of feature vectors are then modeled by a GMM, assuming that frames are independent. GMM-based models for speaker recognition are trained via maximum likelihood (ML) using training data from the target speaker. In general, they need a high number of Gaussian components in order to properly represent the feature space and, therefore, they have a large number of free parameters to be estimated. This makes models prone to overfit when not enough speech from the target speaker is available, which is usually the case.

In order to build systems robust against target speaker data scarcity, a GMM known as UBM was introduced. This way, models known as GMM-UBM emerged, in which the UBM is adapted to the target speaker using the target speech data via maximum a posteriori (MAP) [2]. This UBM is trained with available data from a number of speakers representing the reference population and normalizes the score from the specific target speaker against the rest of the speakers in the dataset.

Moreover, using this GMM-UBM scheme, the information of an utterance can be summarized in a *supervector* of means of the Gaussian components of the model, which provides a fixed-length representation of the segment that simplifies posterior evaluation (scoring). However, this *supervector* is embedded in a high-dimensional space, since typically GMMs of 1024 or 2048 Gaussian components are used, and the feature vector space can be around 60-dimensional (e.g., 20 MFCCs augmented with delta and double delta coefficients). This results in a *supervector* of more than 60k dimensions, which complicates the modeling, comparison and computation of distances between utterances, from where we want to extract similarities and differences to recognize the speaker.

In order to cope with this challenge, techniques based on factor analysis (FA) arose in speaker recognition [4–6]. FA relies on the idea of existing subspaces where channel and speaker variability subspaces are embedded. Then, speaker's identity information and channel information are represented by the latent factors of the two subspaces, which reduces the dimension of the original supervector and provides an easier way to compare utterances, focusing on the target information. However, in order to properly estimate these two subspaces, large training datasets with rich information about variability are required. Moreover, the channel subspace was shown to also contain speaker information [7].

Therefore, the total variability subspace model emerged under the idea of a single subspace where channel and speaker variability (and any other source of variability) are together [3]. The point estimation of the latent factors of the total variability subspaces is known as *i-vector*, and these models have been the state-of-the-art in speaker recognition for several years [3]. Furthermore, given its success, the i-vector model was adapted to perform language recognition [8], becoming a state-of-the-art technique in that field as well.

The main steps of each part of a classical i-vector system for speaker recognition are briefly described below:

- Construction of the UBM (GMM)

    After feature extraction, the signal is represented as a set of feature vectors that are modeled through a generative model as a GMM, i.e., a weighted combination of multi-variate Gaussian components. In this case, the GMM is meant to represent the observed features from several speakers, which are taken as observations from the underlying model. Then, the model is usually trained via expectation maximization (EM) over a large set of utterances that belong to different speakers in different acoustic conditions. The resulting GMM is known as the UBM and represents the spekaer-independent model.

    The UBM obtained is defined by its mean vector (μ, concatenation of mean vectors of each Gaussian component known as *supervector*), its covariance matrix (Σ, covariance matrices of each Gaussian component) and its weight vector (w).

- Baum–Welch statistics estimation

    After the UBM defined by its parameters $\lambda = \{w, \mu, \Sigma\}$ has been trained, the Baum–Welch statistics of each audio segment are computed. For each Gaussian component c, and each utterance frame ut, the zero- and first-order sufficient statistics are as follows:

$$N_c = \sum_t P(c|u_t, \lambda) \tag{2.2}$$

$$F_c = \sum_t P(c|u_t, \lambda)(u_t - \mu) \tag{2.3}$$

where $F_c = \sum_t P(c|u_t, \lambda)(u_t - \mu)$ is the posterior probability of component c generating the frame ut, given by the GMM:

$$P(c|u_t, \lambda) = \frac{w_c N(u_t|\mu_c, \Sigma_c)}{\sum_{j=1}^K w_j N(u_t|\mu_j, \Sigma_j)} \tag{2.4}$$

with N being the Gaussian multi-variate distribution with mean μ_c and covariance matrix Σ_c, and K the number of Gaussian components.

- Total variability model: i-vector

  The underlying idea of the total variability approach is to represent both channel and speaker variability of a given set of training *supervectors* in a low-dimensional subspace T [3]. This matrix T is then a projection matrix trained via EM, with data that ideally includes different sources of variability useful for the target task of speaker recognition in our case.

  We can represent the total variability model as

  $$\mu_u = \mu_{UBM} + Tw \tag{2.5}$$

  where μ_u is the mean supervector representing utterance u, μ_{UBM} is the speaker-independent UBM supervector of means and w is a latent factor, whose point estimate (usually estimated via MAP) is known as the *i-vector*.

  Thus, once the subspace T is trained, the representation given by the i-vector leads to a fixed-length low-dimensional vector at utterance level. This way, in order to extract the i-vector for a given utterance, its Baum–Welch sufficient statistics are collected using the UBM and i-vectors are computed as shown in [3].

  $$w = (I + T^t \Sigma^{-1} NT)^{-1} T^t \Sigma^{-1} F \tag{2.6}$$

  where N and F are matrices composed of the zero- and first-order statistics, and Σ is the covariance matrix of F.

  The resulting i-vectors contain both information of the speaker identity and channel variability (any other sources of nuisance) of the audio segment. The discriminative power and the quality of the information the i-vector is able to capture depends on the variability available in the training dataset used to estimate the total variability subspace (T matrix).

- Scoring

  In i-vector- based systems, this can be done by simply computing a distance measure between i-vectors such as cosine distance. However, typically some techniques are used to compensate and/or account for the variability existing in the i-vector related to the channel and not to the speaker identity. Among the most popular approaches, we can highlight linear discriminant analysis or, the most common in the field, probabilistic linear discriminant analysis (PLDA) [9]. The latter can be intuitively seen as a technique whose aim is to separate the speaker-identity-related information contained in the i-vector from any other information coming from variability sources (channel variability). This is similar to what FA modeling does but on top of low-dimensional i-vectors instead of *supervectors*. Once the subspaces are estimated, PLDA provides a model for efficient scoring of pairs of i-vectors, with different forms of PLDA existing in the literature [10].

- After i-vector computation, the last step of a speaker recognition or verification system is to perform a scoring stage to compare audio segments and determine whether they belong or not to the same speaker.

2.3.2   DNN approaches

2.3.2.1   Basic concepts of neural networks

Neural networks, sometimes also called artificial neural networks, are algorithms for machine learning that consist of a set of units (*neurons*), which are usually grouped in layers (known as *hidden layers*) and connected by some *weights*. This topology allows the model to transform the input data in such a way that it is able to find useful patterns to perform a specific task such as classification or regression. The process of adjusting the values of the connections (weights) is known as *training*.

The currently very popular DNNs are then a type of neural network whose topology is composed of several hidden layers, in contrast to *shallow* architectures with just one hidden layer. With that configuration, DNNs are able to extract features at different levels of abstraction and they also have the ability to learn complex nonlinear transformations of the input directly from the available dataset for the target task. In Figure 2.1, an example of a graphical representation of a DNN is depicted.

DNNs whose neurons in one layer perform a transformation of a weighted sum of the output values of neurons in the preceding layer, as shown in Figure 2.1, are usually known as *feed-forward* or *fully connected* DNNs.

We can summarize the parts of a generic architecture of a DNN used for a classification problem as follows. The DNN would consist of an input layer that receives vectors that represent the input data (for instance, the MFCCs vector representing a frame from an input audio segment). Then, two or more hidden layers would be stacked in order to perform a number of consecutive transformations where the output of a given layer is the input to the next one. This way, the model provides a higher level representation of the input data as we move toward the end (right part in the figure) of the structure. Finally, the model would end with an output layer that computes the output values of the DNN, and whose function and configuration would depend on the target task. For the case of classification, for instance, the model could be defined by an output layer with as many units as classes involved in the problem, and each of the output values could be considered the output probability of an input vector that belong to each class. Thus, in this last layer, for supervised tasks, the output predicted by the network is compared to a reference output label, and the cost of that prediction is computed.

Then, during training, the parameters of the DNN (weight matrices $W_{j,\,j-1}$ and bias vectors b_j, with j from 1 up to the number of hidden layers) are adjusted

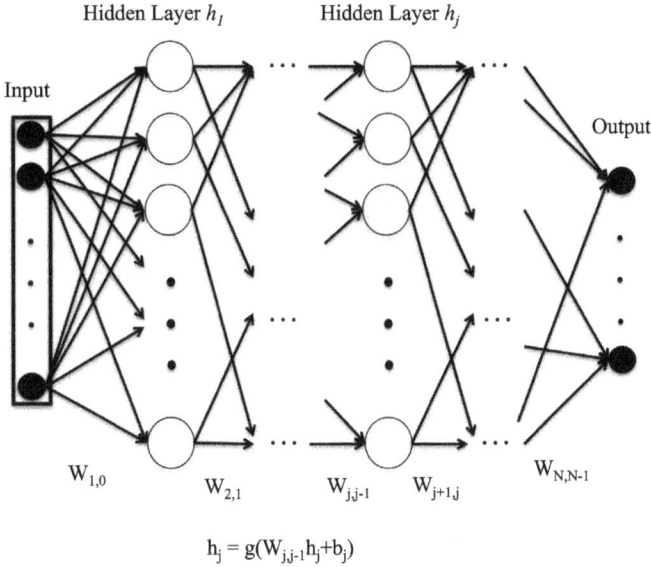

Hidden Layer h_1 Hidden Layer h_j

Input

Output

$$W_{1,0} \qquad W_{2,1} \qquad W_{j,j-1} \quad W_{j+1,j} \qquad W_{N,N-1}$$

$$h_j = g(W_{j,j-1}h_j + b_j)$$

Figure 2.1   *This figure represents a standard feed-forward DNN architecture, with an input layer, a given number of hidden layers and the output layer. Each hidden unit applies a transformation g (usually non-linear function) to the weighted sum of the inputs to that given unit according to the following expression: ($h_j = g(W_{j,j-1}h_j + b_j)$), where W and b are the parameters of the DNN (weights matrices and bias vectors, respectively). Finally, the output layer provides the score vector according to the target task (e.g., for the case of classification, the probability of an input vector belong to each class).*

typically with stochastic gradient descent or some of its variants, based on *back-propagation* of the errors, in order to optimize the cost function defined for the task.

In order to train the network in a supervised way, a training set of sample pairs $(x^{(i)}, y^{(i)})$ should be available, where $x^{(i)}$ is a given input feature vector and $y^{(i)}$ its corresponding label (ground truth). Then, the DNN is fed with the training samples (usually in small *batches* after which the update of the parameters is done), and each hidden layer transforms the data applying generally a non-linear function g. This non-linear function is applied to a weighted linear combination of the output of the previous hidden layer, according to the parameters of the model (weights W and bias b), providing the activation values for each hidden unit as follows for each hidden layer j:

$$h_j(x^{(i)}) = g(W_{j,j-1}h_{j-1}(x^{(i)}) + b_j), \quad j = 2, ..., N - 1 \tag{2.7}$$

$$h_1(x^{(i)}) = g(W_{0,1}x^{(i)} + b_1) \tag{2.8}$$

For the case of classification, usually the output layer is configured to apply a *soft-max* function that estimates the probability $P(c|h(x))$ of a given input x to belong to a certain class c, given by the following equation:

$$P(c|h(x)) = \frac{\exp(W_l^c h_l(x) + b_l^c)}{\sum_{k=1}^{K} \exp(W_l^k h_l(x) + b_l^k)} \tag{2.9}$$

where $h_l(x)$ is the activation of the hidden layer immediately previous to the output layer for a sample x; W_l^c and b_l^c, respectively, refer to the weights matrix and bias vector connecting the output unit for class c with the last hidden layer, and K is the total number of classes.

As mentioned before, the parameters of the model are adjusted in order to mini-mize a cost function. This function represents the cost of obtaining a prediction from the DNN when a given class is expected. The parameters are updated iteratively by means of backpropagation [11], using gradient descent.

The training steps and *feed-forward* architecture described above are part of the *supervised learning* algorithms as opposite to *unsupervised learning* approaches. The former use ground truth labels to estimate the parameters of the model by mini-mizing a cost function that depends on them, while unsupervised algorithms do not make use of any prior information about classes or true labels. Unsupervised tech-niques usually aim to group the data or estimate the distribution that generated them, and are also known as *clustering*.

Due to their use in speaker recognition, we would like to briefly describe the main ideas under two topologies: convolutional neural networks [12] and long short-term memory recurrent neural networks (LSTM-RNNs).

First, convolutional neural networks are very popular models with a slightly different topology from the previously described DNNs. These models have been applied successfully for speaker recognition [13] and have shown their ability to model speech signals for other tasks as well, such as speech or language recognition [14, 15]. They provide a relatively low computational cost algorithm with respect to other DNNs, since their configuration usually reduces the number of parameters of the model.

The topology of convolutional networks is based on a number of filters that are connected through some weights to a region of the two-dimensional input that they receive, and convolved with that region in order to extract features from the input. Moreover, weights are shared among hidden units forming a *feature map,* which in this way extracts the same feature from different locations in the input. This con-figuration reduces the number of parameters to train and, furthermore, they usually perform some *pooling* mechanism too, which further reduces the parameters (size) of the model. All these properties made them very popular, also showing their ability to perform well on noisy acoustic environments.

On the other hand, RNN algorithms introduce connections between units in the same hidden layer, taking as input the output at previous time step, which makes them especially suited for temporal signal modeling. However, RNNs have posed some challenges in order to be properly trained, due to issues such as the *vanish-ing gradient* problem when trying to capture long-term dependencies. One type of

RNNs that solved this problem is the so-called LSTM recurrent networks, which replaced each unit with a *cell*. These cells are controlled by *gates* and store information from previous inputs, being able to keep it during long time periods [16–18], making them suitable models for sequential data by including the information given by the context in a given frame.

2.3.2.2 Some applications of DNNs to speech processing

Part of the motivation to apply neural networks to speaker recognition comes from their success in speech processing and, in particular, in automatic speech recognition (ASR), whose aim is to transcribe the content in a given audio recording.

For many years, ASR systems have focused on algorithms that combine hidden Markov models (HMMs) and GMMs to model both acoustic and temporal information of the audio signal. First attempts to replace the well-established HMM-GMMs using *shallow* neural networks were not able to outperform existing models for many years [19, 20]. However, with the improvements in hardware and the availability of larger datasets, deep architectures started to provide good performance for ASR, which has remarkably progressed in the last decades.

As a first approach, DNNs replaced GMMs for the computation of posterior probabilities of phonetic units in classical HMM-GMM acoustic models, providing better performance than GMMs for that task [21]. These models are known as *hybrid* DNN-HMMs.

Since then, DNNs have been successfully applied in a wide range of manners. For this ASR task, also convolutional neural networks were able to improve these hybrid DNN-HMM models as presented in [14]. Moreover, *end-to-end* approaches in which the whole system is based on DNNs have also shown the ability to succesfully perform ASR, providing more independence from hand-designed components. One example is the system based on recurrent networks (RNNs) described in [22], which deals with ASR in noisy environments. Some more works based on DNNs for end-to-end ASR studied their ability to perform the task in multilingual frameworks and in real time [23], or as *bottleneck* feature extractors instead of end-to-end for continuous ASR [24–26].

2.3.3 DNNs for speaker recognition

As in many other areas, DNNs have irrupted in the speaker recognition field, becoming state-of-the-art techniques. The introduction in the field is strongly motivated by their success in speech recognition. Broadly speaking, we can say that they have been mainly used in four different ways: replacing the UBM when accumulating statistics in the i-vector pipeline, as feature extractors, as *embedding* or *x-vector* extractors and as end-to-end systems. We will briefly present each of them.

- DNNs for statistics estimation
  As it happened in other areas, DNNs started to replace parts of classical approaches, as the pipeline based on i-vectors, in the speaker recognition field.

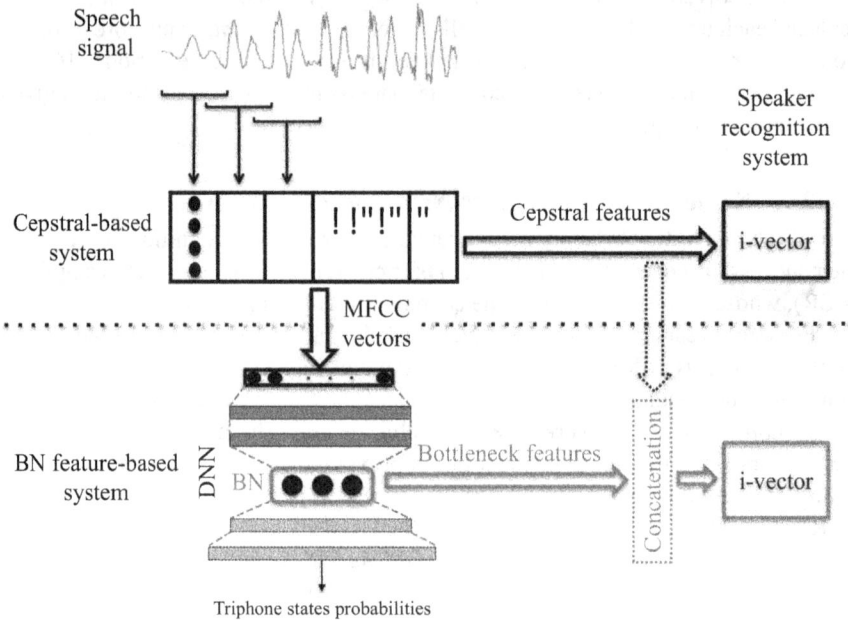

Figure 2.2   *This figure shows a graphical representation of a cepstral (upper part) versus bottleneck (bottom part) feature-based speaker recognition system. For the bottleneck (BN) feature-based approach, sometimes, MFCCs are concatenated with the bottleneck features before the i-vector modeling.*

This way, the DNNs are used to replace the UBM (GMM) when accumulating statistics in such a way that now it is the DNN that provides posterior probabilities for frame alignments instead of the UBM as it is done in works such as [27, 28].

This approach can be enriched by using different features for frame alignments and statistics estimation, as it is done in [28]. These *hybrid* systems are able to leverage the potential of both features designed for speaker discrimination and optimal features for frame alignments [29].

• *Bottleneck* features

One of the most successful approaches to language recognition is the use of DNNs as feature extractors, which extract frame-by-frame vectors learned directly by the DNN. This approach has also been used in speaker recognition [30–32], usually using both information from acoustic features and *bottleneck* features as represented in Figure 2.2.

For these systems, the DNN is trained to discriminate among a given set of *senones* or phoneme states as in the previous case. However, the topology of the DNN is a *feed-forward* network that includes a *bottleneck* layer. This is a relatively small hidden layer with respect to the rest, which aims to capture and compress the information learned by the DNN up to that point. Since the network is usually trained for ASR, these *bottleneck* layer activations can be seen as a projection of the transformation of the input feature vectors provided by the network into a new representation of the signal that contains information useful for the task of ASR [24]. In other words, *bottleneck* features can be described as a new frame-wise representation of an audio segment that provides more abstract feature vectors that model the feature space and allows the network to learn them by itself. This also reduces the dependence on hand-crafted features.

We should note that training the DNN for ASR in other areas such as language recognition is motivated by the fact that phoneme sequences are different for each language [30, 33]. However, in speaker recognition, although the DNN has proven its ability to capture phonetic information that helps characterizing different speakers [30–32], performance of stand-alone *bottleneck* features is usually improved by the use of joint acoustic (MFCCs, PLPs) and *bottleneck* features in the i-vector/PLDA pipeline [31], obtaining more discriminative feature vectors.

- *Embeddings* or *x-vectors*

  As described in Section 2.4, DNNs are started to be explored in text-dependent speaker recognition to replace both GMM-UBM and i-vector models, and used directly to extract a segment level representation known as *embedding* [34, 35].

  An *embedding* provides a compact fixed-length representation of an utterance of variable duration, as opposed to frame-wise feature vectors such as classical acoustic features (MFCCs or PLPs) or *bottleneck* features. In this sense, they are similar to i-vectors, but *embeddings* are learned by a DNN usually trained discriminatively to classify a given set of speakers, so instead of capturing information about the channel and speaker variability, they are meant to compress information relevant to the speaker recognition task and then represent the speaker's identity features.

  For text-independent speaker recognition, *embeddings*, also called *x-vectors*, obtained from a fairly simple DNN topology were presented in [36]. In this type of systems, the *embeddings* are computed as the output values of a hidden layer that works at segment level, after some summary of the information at frame-level has been performed.

  As represented in Figure 2.3, the DNN has two parts separated by the summarizing or *pooling* layer. The first part that goes from the input up to the pooling layer works on a frame-by-frame basis. This way, the input layer is fed by a sequence of feature vectors such as MFCCs and is followed by a number of hidden layers that ideally aim to capture the information contained in this frame-by-frame representation of the input, with its temporal context. This could be done, for instance, stacking some LSTM or time-delay neural network layers. Then, a *pooling* layer summarizes the frame-by-frame output of the previous layer by, for example, computing mean and standard deviation over time, providing a single vector of values for each sequence. Finally, the second part of the DNN, which connects the pooling layer with the output layer, can be composed of one

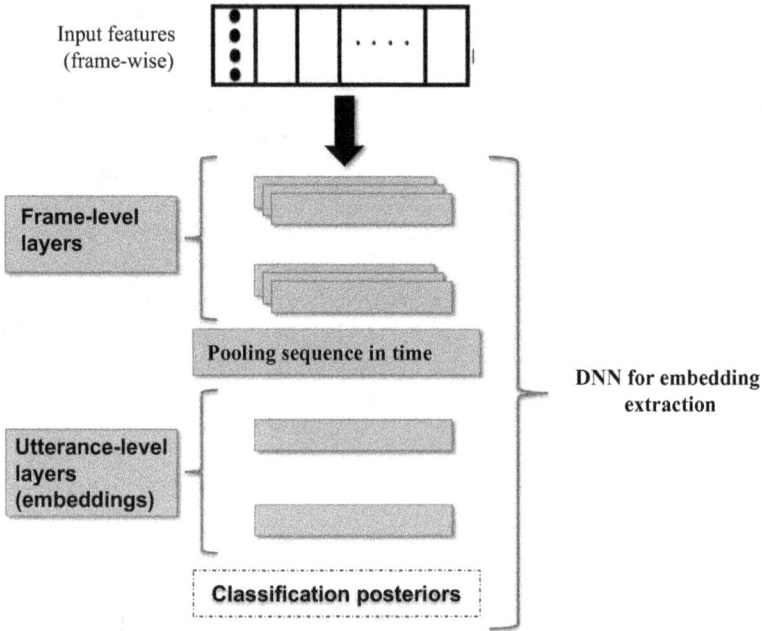

Figure 2.3   This is a representation of the generic scheme of a DNN used as embedding extractor.

or more *fully connected* hidden layers that act over segments instead of frames. The activation values of these layers can be considered a new representation of the input segment: the *embedding*.

The DNN is trained to classify a given set of speaker, and thus, the output layer provides the posterior probabilities of each speaker in the training set. The system is trained as a whole with typically gradient descent optimizer and cross-entropy cost function.

After training, the output layer is discarded and the DNN is used to forward new input sequences through the network up to the *embedding* layers in order to extract a fixed-length representation of the segment. The resulting *embeddings* or *x-vectors* are usually modeled with classical backends for i-vectors such as PLDA [37].

These systems are widely explored in the field since *embeddings* provide comparable or better performance than well-established i-vectors [36], especially when we deal with short utterances and there is a large enough dataset available for training.

- *End-to-end* DNNs for speaker recognition
- Even though speaker verification (the most common task of speaker recognition) is a challenging scenario still to be addressed by DNNs in an *end-to-end* fashion, i.e., which provides the final output score for each trial, some attempts have been done. As presented in Section 2.4, successful results for text-dependent

speaker recognition have been achieved when using a DNN trained for speaker classification as embedding (*d-vector*) extractor and performing the target task with another DNN that outputs a score for each trial. However, for text-independent speaker recognition, the exploration of this approach is still in an early stage, with some promising results presented in [38, 39], which has drawn the attention of many researchers nowadays.

2.4 Text-dependent speaker recognition

This chapter is more focused in text-independent speaker recognition, where the lexical content of the speech is unknown and therefore the techniques used try to minimize the influence of the lexical content in the recognition. However, there are a number of possible applications of speaker recognition in which the speaker actively cooperates with the system to perform the authentication (for instance, in voice authentication applications where the user tries to get identified to access a private information such as in banking applications). In these systems, it is possible to know the lexical content of the utterance used for verification, either because it is a fixed phrase, a user-specific pass-phrase or even a random phrase prompted by the system to the user. This knowledge can be exploited by the system with different purposes. First, it can be used to more accurately model the acoustic sequence, which allows better accuracy even for very short utterances. Second, it can be used to verify that not only the acoustics match the speaker characteristics, but also that the lexical content is the expected lexical content (adding a security level using a PIN code or avoiding replay attacks by prompting random pass-phrases). The speaker recognition systems that exploit this knowledge of the lexical content of the utterance to verify are called text-dependent speaker recognition systems. This difference between text-dependent and text-independent systems can make these two types of speaker recognition systems very different in terms both of techniques used and of potential applications. This section is devoted to text-dependent speaker recognition systems. In particular, this section will present a brief review of the techniques and the main databases and benchmarks in this field.

2.4.1 Classification of systems and techniques

We can classify text-dependent speaker recognition systems from an application point of view into two types: fixed-text and variable-text systems. In fixed-text systems, the lexical content in the enrollment and the recognition samples is always the same. In variable-text systems, the lexical content in the recognition sample may be different in every access trial from the lexical content of the enrollment samples. Variable-text systems are more flexible and more robust against attacks that use recordings from an user or imitations after hearing of the true speaker uttering the correct pass-phrase.

With respect to the techniques used for text-dependent speaker recognition, it has been demonstrated [40] that information present at different levels of the speech signal (glottal excitation, spectral and suprasegmental features) can be used effectively to detect the user's identity. However, the most widely used information is the spectral content of the speech signal, determined by the physical configuration and dynamics of the vocal tract. This information is typically summarized as a temporal sequence of MFCC or filter bank output vectors, each of which represents a window of 20–40 ms of speech. In this way, the problem of text-dependent speaker recognition is reduced to a problem of comparing a sequence of feature vectors to a model of the user. For this comparison there are mainly three methods that have been used: template-based, statistical and deep learning methods. In template-based methods [41, 42], the model of the speaker consists of several sequences of vectors corresponding to the enrollment utterances, and recognition is performed by comparing the verification utterance against the enrollment utterances using dynamic time warping (DTW). These were the first methods applied in this field but still may find application, particularly for embedded systems with very limited resources. Statistical methods, mainly based in HMMs [43], provide more flexibility, allow to choose different speech units (from sub-phoneme units to words) and enable the design of text-prompted systems [44, 45] that are robust against replay attacks. HMMs can be used to perform speech recognition and (using speaker adaptation techniques) verify the speaker as well. With the advent of deep learning in the last few years, there has been a huge interest in exploring the use of DNNs in text-dependent speaker recognition. One of the first successful approaches was based in obtaining an utterance-level representation, called *embedding*. This embedding is obtained by computing the mean over the framewise outputs of one or more layers in the DNN in [34] or by the use of an RNN in [35]. These systems mixed the DNNs with other techniques. More recently, particular interest has been placed on end-to-end DNN systems where a whole task is performed using only DNNs. In the case of text-dependent speaker recognition, some successful end-to-end DNN-based approaches have already been proposed [34, 35], but still using a DNN previously trained for speaker classification in order to extract embeddings (*d-vectors*) and performing the backend with another DNN that outputs the scoring for each trial.

2.4.2   *Databases and benchmarks*

The first databases used for text-dependent speaker verification were databases not specifically designed for this task, like the TI-DIGITS [46] and TIMIT [47]. One of the first databases specifically designed for text-dependent speaker recognition research is YOHO [48]. This was the most extended and well-known benchmark for comparison for many years and was frequently used to assess text-dependent systems. The YOHO database had several limitations, such as being recorded with a single microphone in a quiet environment and not designed to simulate informed forgeries (i.e., impostors uttering the password of an user). Databases assessing these limitations were proposed, such as the MIT Mobile Device Speaker Verification Corpus [49], designed to allow research on text-dependent speaker verification on realistic noisy conditions, and the BIOSEC Baseline Corpus [50], designed to simulate informed forgeries. Other benchmarks have been produced such as the data used in the MOBIO (Mobile Biometrics) evaluation [51], the RedDots dataset [52] and the RSR2015 corpus [53, 54].

One of the main difficulties of the comparison of different text-dependent speaker verification systems is that these systems are usually language-dependent, and therefore many researchers tend to present their results in their custom and typically proprietary database, making it impossible to make direct comparisons. The comparison of different commercial systems is even more difficult. Besides, as with other biometric modalities, technical performance is not the only dimension to evaluate and other measures related to the usability of the systems should be evaluated as well [55].

More recently, the DeepMine dataset [56] has been released, which is the largest public text-dependent and text-prompted speaker verification database containing two languages: Persian and English. It also contains data for speech recognition and text-independent speaker recognition in Persian. It contains more than 1 850 speakers and 540 thousand recordings, with more than 480 hours of transcribed speech. This corpus has been used to organize the Short-duration Speaker Verification challenges 2020 [57] and 2021 [], providing a new common benchmark in the field.

2.5   Calibration of speaker recognition scores

2.5.1   Motivation: why to calibrate?

So far, we have described ways of obtaining system scores that can be as discriminating as possible. Discrimination is understood as the ability of system scores to distinguish comparisons where the speakers in both speech materials come from the same source (i.e., the same-source hypothesis), from comparisons where the speech materials come from different sources (i.e., the different-source hypothesis). This *separation* can be performed by the establishment of a threshold, in order to minimize some cost function depending on the false acceptances, the false rejections and the cost of their consequences. Throughout this chapter, we are assuming that system scores support the same-source hypothesis more strongly when its value will be higher, and lower scores support the different-source hypothesis more strongly. This assumption does not imply a loss of generality, and analogous reasoning can be followed assuming the contrary. Discrimination is typically measured classically with DET/ROC curves, the area under a ROC curve or the EER.

However, in some applications, this discrimination property of system scores is neither enough nor convenient on its own. For instance, if we have no previous knowledge about a system, a system score in itself means nothing, since it is just a number that cannot be compared to any reference. It is only by the knowledge about other scores of that particular system that a single score becomes meaningful. In particular, in the absence of other additional knowledge about the system, we can only establish whether one of its scores is supporting same-source or different-source hypotheses in a given comparison, and with what strength, if we know how scores range when the same-source hypothesis is true (i.e., same-source scores), and how do they behave if the different-source hypothesis is true (i.e., different-source scores). Therefore, one cannot evaluate a single score from the system without a previous study of the properties of the system scores, which for some operational scenarios can be burdensome or not appropriate (e.g., commercial off-the-shelf systems that the customer is willing to use in a straightforward manner and without previous experimentation).

Another motivation for the calibration of system scores is found in the context of the decision scenario where the system will be used. For a given application, the system will be used to ultimately make a decision. The conditions and constraints on that decision are generically included in the so-called *application policy*. For instance, in a high-security scenario (e.g., access to a bank account), optimal decisions could be obtained with higher decision thresholds, since false acceptance decisions are considered more harmful. On the other hand, high-usability scenarios (e.g., easy access to recommended publicity by a user) require false rejections to be minimal, and therefore lower thresholds are desirable. According to Bayesian decision theory, it is not only the consequence of a decision what matters to establish a threshold according to a policy, but also the prior probability of a same-source or a different-source attempt. This follows the logic that, if there are very unlikely same-source comparisons, it is less probable that false acceptances will happen, and therefore we can reduce the threshold accordingly to satisfy the decision policy.

A special case of this formal, policy-driven decision scheme is forensic voice comparison, where the system is used to make decisions in a criminal trial [58]. Here, the speech from a suspect is compared to an incriminating set of speech materials of unknown origin, in the context of a criminal offense. Thus, using the information of the forensic comparison of speech materials, a fact finder (a judge or a jury) will make the final decision as whether the suspect must be condemned or acquitted. In this context, the application policy with respect to the consequences of wrong decisions may be established for a given trial in advance, and there might be even some guidelines about it. For instance, in most Western, democratic countries, false acceptances are typically considered much worse than false rejections, following the principle of the presumption of innocence. However, the prior probability of the suspect being the author of the incriminating recording (i.e., the prior same-source probability without the consideration of the speech evidence) varies from case to case, and it is related with all the possible circumstances in the case. For instance, in a given case, the prior probability is seriously affected by police investigations, the presence of witnesses, the criminal record of the suspect, the previous relationship of the suspect with the particular crime, etc. Hence, the application policy is particular for each given case, and it will very likely vary from one case to another.

In this application policy-driven decision framework, it seems reasonable that the speaker recognition system might yield some kind of score that can be used to make optimal decisions without the need of re-generating system scores, or use a previously generated set of scores, to re-tune the threshold each time the application policy changes. This goal is accomplished by the calibration of system scores.

2.5.2   *What is calibration?*

Calibration is a property of system scores. According to it, scores that are well cali-brated can be used to make optimal decisions following Bayesian decision theory, provided that an application policy has been previously established.

Formally, according to Bayesian decision theory, in the comparison of two sets of speech materials, namely x_1 and x_2, the system computes a so-called observed score s, i.e., an instance of a random variable S representing the scores output by the system. In this context, the unobserved variable of interest is the true binary hypothesis H, with possible values h_1 and h_2, mutually exclusive and exhaustive. These hypotheses are typically defined in speaker recognition in the following way:

h_1: x_1 and x_2 were originated by the same speaker (same-speaker hypothesis).

h_2: x_1 and x_2 were generated from different speakers (different-speaker hypothesis).

In the context of speaker recognition for commercial applications, this hypoth-esis definition is typically accepted as adequate in general. However, in forensic contexts, other propositions can be addressed, and variation of their statement can lead to radically different score calibration schemes. Therefore, care should be taken in order to clearly and appropriately define the propositions in a case. Literature in the topic can be found in [59–61].

Following Bayesian decision theory, and according to the previous assump-tions, we have the following expression:

$$\frac{P(h_1|s)}{P(h_2|s)} = \frac{p(s|h_1)}{p(s|H_2)} \times \frac{P(h_1)}{P(h_2)} \tag{2.10}$$

namely the odds-form of Bayes' theorem, where $P(h_1|s)$ is the so-called posterior probability of the proposition h_1 given the observed score, representing the probabil-ity that the observed score will be generated by a same-speaker comparison. This posterior probability depends on the so-called likelihoods, which are probability densities particularized to the value of s and conditioned to each of the hypotheses, namely $p(s|h_1)$ and $p(s|h_2)$[1], and also on the prior probabilities $P(h_1) = 1 - P(h_2)$. Thus, the posterior probabilities represent our knowledge (uncertainty) about the true value of the hypothesis H (i.e., our variable of interest) once the score is observed. On the other hand, the prior probability represents our knowledge (uncer-tainty) about the true value of the hypothesis H *before* the score is observed. The contribution of the system score to this decision problem is given by the so-called likelihood ratio (LR):

$$LR = \frac{p(s|h_1)}{p(s|h_2)} \tag{2.11}$$

The aim of a calibration method is tonobtain the LR from the observed score s, in a so-called score-to-LR, or calibration, transformation. This way, if the LR is

---

[1]Note that those likelihoods are not complementary probabilities, and not even probabilities, but prob-ability densities.

computed in a correct way (i.e., if the probabilities in (2.11) are well represented), then optimal solutions can be made by the use of the well-known Bayes decision rule, namely:

$$\text{Decide } h_1 \quad : \frac{p(s|h_1)}{p(s|h_2)} > \frac{p(h_2)}{p(h_1)} \frac{C_{fa}}{C_{fr}} \tag{2.12}$$

$$\text{Decide } h_2 \quad : \text{Otherwise} \tag{2.13}$$

where C_{fa} is the cost of the consequence of a false acceptance (i.e., the cost of deciding h_1 in a comparison where h_2 is true), and C_{fr} is the cost of the consequence of a false rejection (i.e., the cost of deciding h_2 where h_1 is true). Thus, the prior probabilities $P(h_1)$ and $P(h_2)$, together with the decision costs, C_{fa} and C_{fr}, define the application policy, being able to range from high-security to high-usability scenarios as mentioned before.

Moreover, it can be seen from (2.13) that the separation between the system and the application policy is clear. The system contributes with the LR, and the application policy is set by the prior probabilities and the costs of the decision consequences. Therefore, if the application policy changes, and the system outputs an LR value, the threshold can be immediately set to keep obtaining optimal decisions, without the need of re-generating system scores, or use existing ones, to re-tune an *ad-hoc* decision threshold. This property overcomes all the difficulties for optimal decision-making described in Section 2.5.1.

As a matter of fact, by the computation of an LR from an observed score, what is really being conducted is a transformation between a meaningless score and a meaningful LR value. Thus, the LR in itself has a meaning, i.e., "it is LR times more likely to observe system score s if both speech materials x_1 and x_2 were originated by the same speaker (h_1 is true), that if they were originated by different speakers (h_2 is true)." This has strong implications in areas such as forensic science, as mentioned section 2.5 [58, 62].

Summarizing, well-calibrated LR values will be computed either by assigning appropriate values of the likelihoods separately or by assigning values to the LR in the decision problem as a whole. Hence, using well-calibrated LR values to make Bayes decisions with (2.13) for a given application policy will yield close-to-minimum decision costs, and therefore they will be almost optimal. On the other hand, if the probabilities involved in (2.11) are computed badly or under bad assumptions, or if the calibration scheme does not take into account the behavior of the LR, the calibration of the resulting LR values will be defective, and the decisions will not be close to the optimal Bayes decisions anymore, leading to higher decision costs.

As a final remark, it has been stated that calibration is a property of a set of scores, since it measures the ability of a set of scores to work well in terms of decision costs when Bayes decisions are made (2.13). Also, in order to calibrate, we need to compute an LR value, and we also talk about the calibration of LR values. Although this might seem confusing, it makes all sense, since it turns out that an LR value is indeed a score. The only needed property of a score is that it discriminates comparisons where h_1 and h_2 are true. But an LR value, as defined here, has that property indeed, according to Bayesian decision theory. Therefore, calibration is indistinctly a property of a set of scores and a property of a set of LR values. The opposite is not true: any score cannot be said to be an LR value in general, especially if (2.11) was not taken into consideration, or in other words, if the computation of the LR value did not take into account any probabilistic model. As a consequence, a score that was not computed as an LR might present bad calibration, and if it presents good calibration, it will be most possibly by chance.

2.5.3 Score-to-LR computation methods

The main commonality of all the methods described in this section is that they need two hypothesis-dependent sets of *training* scores, namely S_1 and S_2. These sets and some of the issues associated with them are described as follows.

- The set $S_1 = \left\{ s_1^{(1)}, ..., s_1^{(N_1)} \right\}$ consists of a number N_1 of scores computed assuming that $H = h_1$. Therefore, the selection of speech materials to compute the scores in S_1 has to be done accordingly to the definition of the hypothesis h_1. As the h_1 proposition typically assumes that x_1 and x_2 come from the same person, the S_1 consists of *same-source* scores. However, the rest of information in h_1 can be determinant in order to select the database to generate those same-source scores. For instance, if the particular speaker in a comparison is known, as, i.e., the suspect in a forensic case where speech is involved, the hypothesis definition could include this identity (e.g., "the trace was left by Mr Dean Keaton"), and then the propositions would be *person-specific* or *source-specific*. In that case, the database to generate S_1 should include speech coming from the particular suspect, because in speaker recognition each person has a particular behavior regarding their score distribution [63]. On the other hand, *person-generic* or *source-generic* propositions (e.g., "the trace and the reference samples come from the same source") would allow the use of any same-source score from other people, since there is not a reference from a particular subject. The latter has been dubbed *the common source problem* in the forensic literature [61].

- The set $S_2 = \left\{ s_2^{(1)}, ..., s_2^{(N_2)} \right\}$ consists of a number N_2 of scores computed assuming that $H = h_2$. For the computation of these scores, several things should be taken into account. As h_2 typically assumes that the x_1 and x_2 speech materials were not generated by the same person, the scores in S_2 will essentially be *different-source* scores. Also, issues about hypotheses definition and data selection apply here, and are of special importance in forensics [61, 64].

The training scores in S_1 and S_2 should represent all possible sources of variability in generating score random variable S. Therefore, the use of models of variability compensation in speaker recognition is essential in order to compute better LRs. Fortunately, good examples exist in the speaker recognition literature of the use of variability compensation models at different levels in the system [3, 65]. Moreover, the condition of the training scores must resemble those of the speech to be compared as much as possible, and remarkable work has been done recently in order to automatically select those training scores for calibration following that idea [66, 67].

2.5.3.1   Generative calibration models: fitting distributions to scores

The first attempts of calibration models for speaker recognition scores were proposed in forensic voice comparison by the use of simple, generative techniques modeling the hypotheses-conditional distributions of the score random variable S. This is the approach already presented in [68] and has been followed in subsequent works in the literature.

Under this approach, the objective is assigning the likelihoods $p\left(S|\,h_1\right)$ to the training scores $\mathbf{S_1}$, and $p\left(S|\,h_2\right)$ to $\mathbf{S_2}$. Then, the ratio of the particular value of these densities at $S = s$ will be the LR value.

Assigning $p\left(S|\,h_1\right)$ and $p\left(S|\,h_2\right)$ implies the selection of a proper model. The most straightforward choice for speaker recognition scores could be the Gaussian distribution. Score normalization techniques might help to suit the scores to a given distribution such as a single Gaussian [69], but the good fitting to a Gaussian must be checked in advance, otherwise the calibration can be poor.

Some approaches for generative density fitting includes the use of kernel density functions [68, 70], Gaussian mixture models [70] and other parametric distributions (see examples in automatic fingerprint recognition such as [71, 72]), all of them following a maximum likelihood (ML) approach. Speaker-dependent approaches have been followed in [73, 74], where the speaker information is exploited in person-specific forensic scenarios. Recent works have proposed the use of Bayesian inference, which has proven to work nicely where few scores are available (a typical situation in speaker-dependent calibration strategies). Also, unsupervised generative models have been proposed in [75], where the labels for h_1 and h_2 are not known in the training set, allowing the use of massive amounts of scores in source-generic scenarios. Finally, other distributions have been suggested in [76], yielding nonlinear score-to-log-LR transformations.

2.5.3.2   Discriminative calibration models: transforming scores into LR values to optimize a cost function

The most commonly used model for calibration in speaker recognition is logistic regression, a well-known pattern recognition technique widely used for many problems including fusion [77, 78] and more recently LR computation [79–81]. The aim of logistic regression is obtaining an affine transformation (i.e., shifting and scaling) of an input dataset in order to optimize an objective function. Let $\mathbf{S_f} = \left\{s_f^{(1)}, s_f^{(2)}, \ldots, s_f^{(K)}\right\}$ be a set of scores from K different speaker recognition systems. The affine[2] transformation performed by the logistic regression model can be defined as

$$f_{lr} = \log\left(\frac{P(h_1|S_f)}{P(h_2|S_f)}\right) = a_0 + a_1 \cdot s_f^{(1)} + a_2 \cdot s_f^{(2)} + \cdots + a_K \cdot s_f^{(K)} \tag{2.14}$$

This leads to the following *logistic regression model*:

$$P\left(h_1 | S_f\right) = \frac{1}{1+e^{-f_{lr}}} = \frac{1}{1+e^{-\log(LR)-\log(O(h_1))}} \tag{2.15}$$

where the term $O\left(h_1\right)$ are the prior odds of the hypothesis h_1, namely $\frac{P(h_1)}{P(h_2)}$. The weighting terms $\{a_0, a_1, a_2, \ldots, a_K\}$ can be obtained from the training scores with well-known optimization procedures found in the literature, such as gradient descent or conjugate gradient.[3] Logistic regression typically optimizes the objective function for the value of the *empirical prior* (i.e., the proportion of the number of training scores where h_1 is true over the total number of training scores, or $\frac{N_1}{N_1+N_2}$), but removing the influence of that empirical prior from the obtained posterior odds, LRs are obtained.

Notice that logistic regression can be used not only to compute LRs from a single speaker recognition score ($K = 1$) but also to perform fusion and LR computation simultaneously (when $K > 1$) [78]. This fact, joined to the good behavior that logistic regression presents in most situations, has made this LR computation algorithm one of the most popular ones.

Another approach to score calibration has been proposed by the use of the pool adjacent violators (PAV) algorithm [79]. The PAV algorithm transforms a set of scores into a set of LR values presenting optimal calibration. However, it is only possible to apply an optimal PAV transformation if the ground truth labels of the propositions for each score in the set are known. Nevertheless, as suggested in [82], a PAV transformation can be trained on the set of training scores S_1 and S_2, and then apply the trained transformation to a score in a forensic case. Although a straightforward use of PAV leads to a non-invertible transformation, which is numerically an issue, several *smoothing* techniques can be applied to PAV in order to keep it monotonically increasing. For instance, adding a very small, but yet positive, slope to PAV will lead to an invertible transformation. Interpolating with linear, quadratic or splines approaches are also possible smoothing schemes.

---

[2]Notice that an affine transformation in the domain of the log-probability-ratio, or log-odds, is equivalent to a sigmoid in the probability domain. A sigmoid transformation is perhaps a much more typical form of presenting logistic regression in pattern recognition and machine learning, but in speaker recognition the affine transformation in the log-odds domain is much more typically seen. That is why some speaker recognition literature sometimes use the term linear logistic regression, which causes confusion in other fields.

[3]Typical implementations used in speaker recognition include toolkits like FoCal or BOSARIS, which can be found in http://niko.brummer.googlepages.com.

2.5.4   *Performance measurement of score-to-LR methods*

In speaker recognition, performance measurement is typically done in an empirical way, following remarkable benchmarks such as NIST SREs. Following those, it is possible to obtain a set of N_1 calibrated scores (i.e., LR or log-LR values) computed when h_1 is true and a set of N_2 LR values computed when h_2 is true.

Regarding performance measurement, typical discrimination measures such as DET plots and the EER are not adequate to measure the performance of LR values, because they do not take calibration into account. In particular, two sets of scores might have exactly equal discrimination (DET curve) but very different calibration. For instance, if we add a constant value to a set of scores, the discrimination of the scores before and after summing the constant is exactly the same, but the calibration can be very different if the constant is relatively big.

A solution to measure the performance of LR values has been proposed in [79] for speaker recognition, and has been dubbed *log-likelihood-ratio cost* (C_{llr}). C_{llr} is a scalar measure of performance of LR values, the lower its value the better the performance. Later, it has been used and generalized in many other fields, such as forensic voice comparison and other forensic science disciplines [83–87]. C_{llr} is defined as follows:

$$C_{llr} = \frac{1}{2 \cdot N_1} \sum_{i_1} \log_2 \left(1 + \frac{1}{LR_{i_1}} \right) + \frac{1}{2 \cdot N_2} \sum_{j_2} \log_2 \left(1 + LR_{j_2} \right) \tag{2.16}$$

The indices i_1 and j_2, respectively, denote summing over the LR values of the simulated cases where each proposition h_1 and h_2 is respectively true. An important result is derived in [79], where it is demonstrated that minimizing the value of C_{llr} also encourages to obtain reduced Bayes decision costs for all possible decision costs and prior probabilities [88]. This property has been highlighted as extremely important in forensic science [89]. Moreover, in [79], the PAV algorithm is used in order to decompose C_{llr} as follows:

$$C_{llr} = C_{llr}^{min} + C_{llr}^{cal} \tag{2.17}$$

where

- C_{llr}^{min} represents the *discrimination cost* of the LR method, and it is due to non-perfect discriminating power.
- C_{llr}^{cal} represents the *calibration cost* of the system.

Additionally, a *neutral* value of $C_{llr} = 1$ is obtained when all the LR values computed by the system are equal to 1, meaning a *useless* system that adds nothing to the decision to be made between h_1 and h_2. Thus, this value of $C_{llr} = 1$ represents an ultimate performance limit, and if the system presents a C_{llr} value higher than that, it should not be used for decisions. Finally, interesting information-theoretical interpretations can be drawn from the value of C_{llr}, as its value is strongly related to the quantity of the cross-entropy, well known in the machine learning field [79, 87].

Other graphical representations circumventing these ideas have been proposed in the literature, such as applied probability of error plots [79], normalized Bayes error plots (NBE) and empirical cross-entropy plots [87]. All these measures emphasize the decoupling between the contribution of the system to the decisions (encapsulated in the LR or calibrated score) and the rest of information (encapsulated as the application policy, i.e., prior probabilities of the hypotheses and costs of decision consequences). Recent efforts have tried to relate classical measures of score discrimination, such as DET curves, with the calibration of scores and LR values, in different application-policy scenarios [90].

References

[1] Kinnunen T., Li H. 'An overview of text-independent speaker recognition: from features to supervectors'. *Speech Communication*. 2010;52(1):12–40.

[2] Reynolds D.A., Quatieri T.F., Dunn R.B. 'Speaker verification using adapted Gaussian mixture models'. *Digital Signal Processing*. 2000;10(1):19–41.

[3] Dehak N., Kenny P.J., Dehak R., Dumouchel P., Ouellet P. 'Front-End factor analysis for speaker verification'. *IEEE Transactions on Audio, Speech, and Language Processing*. 2011;19(4):788–98.

[4] Kenny P., Boulianne G., Ouellet P., Dumouchel P. 'Joint factor analysis versus Eigenchannels in speaker recognition'. *IEEE Transactions on Audio, Speech and Language Processing*. 2007;15(4):1435–47.

[5] Kenny P., Boulianne G., Ouellet P., Dumouchel P. 'Speaker and session variability in GMM-based speaker verification'. *IEEE Transactions on Audio, Speech and Language Processing*. 2007;15(4):1448–60.

[6] Kenny P., Ouellet P., Dehak N., Gupta V., Dumouchel P. 'A study of inter-speaker variability in speaker verification'. *IEEE Transactions on Audio, Speech, and Language Processing*. 2008;16(5):980–8.

[7] Dehak N. *Discriminative and Generative Approaches for Long- and Short-Term Speaker Characteristics Modeling: Application to Speaker Verification*. Canada: École de Technologie Superiere, Université du Quebec; 2009. Available from https://espace.etsmtl.ca/id/eprisnt/33/.

[8] Martínez D.G., Plchot O., Burget L. 'Language recognition in iVectors space'. *Proceedings of Interspeech 2011*. 2011;861–4.

[9] Prince S.J.D., Elder J.H. 'Probabilistic linear discriminant analysis for inferences about identity'. *2007 IEEE 11th International Conference on Computer Vision*; 2007. pp. 1–8.

[10] Kenny P. *Bayesian Speaker Verification with Heavy-Tailed Priors*. Odyssey; 2010.

[11] Bishop C.M. *Pattern Recognition and Machine Learning Information Science and Statistics*. Secaucus, NJ, USA: Springer-Verlag New York, Inc; 2006.

[12] LeCun Y., Bottou L., Bengio Y. 'Gradient-based learning applied to document recognition' in Haykin S., Kosko B. (eds.). *Intelligent Signal Processing*. IEEE Press; 2001. pp. 306–51.

[13]   McLaren M., Lei Y., Scheffer N. 'Application of convolutional neural networks to speaker recognition in noisy conditions'. INTERSPEECH 2014, 15th Annual Conference of the International Speech Communication Association; Singapore; 2014. pp. 14–18.

[14]   Abdel-Hamid O., Mohamed A.-rahman., Jiang H., Deng L., Penn G., Yu D. 'Convolutional neural networks for speech recognition'. *IEEE/ACM Transactions on Audio, Speech, and Language Processing.* 2014;22(10):1533–45.

[15]   Lei Y., Ferrer L., Lawson A. 'Application of convolutional neural networks to language identification in noisy conditions'. ODYSSEY; 2014.

[16]   Graves A. *Supervised Sequence Labelling with Recurrent Neural Networks.* Springer; 2012. p. 385.

[17]   Gers F.A., Schmidhuber J., Cummins F. 'Learning to forget: continual prediction with LSTM'. *Neural Computation.* 2000;12(10):2451–71.

[18]   Gers F.A., Schraudolph N.N., Schmidhuber J. 'Learning precise timing with LSTM recurrent networks'. *Journal of Machine Learning Research.* 2003;3:115–43.

[19]   Hinton G., Deng L., Yu D., *et al.* 'Deep neural networks for acoustic modeling in speech recognition'. *Signal Processing Magazine.* 2012.

[20]   Bourlard H.A., Morgan N. *Connectionist Speech Recognition: A Hybrid Approach.* Norwell, MA, USA: Kluwer Academic Publishers; 1993.

[21]   Hinton G., Deng L., Yu D., *et al.* 'Deep neural networks for acoustic modeling in speech recognition: the shared views of four research groups'. *IEEE Signal Processing Magazine.* 2012;29(6):82–97.

[22]   Hannun A.Y., Case C., Casper J., *et al.* Deep speech: Scaling up end-to-end speech recognition. CoRR. 2014. Available from http://arxiv.org/abs/1412.5567.

[23]   Gonzalez-Dominguez J., Eustis D., Lopez-Moreno I., Senior A., Beaufays F., Moreno P.J. 'A real-time end-to-end Multilingual speech recognition architecture'. *IEEE Journal of Selected Topics in Signal Processing.* 2015;9(4):749–59.

[24]   Grezl F., Karafiat M., Kontar S. 'Probabilistic and bottle-neck features for LVCSR of meetings'. 2007 IEEE International Conference on Acoustics, Speech and Signal Processing – ICASSP '07; 2007. pp. 757–60.

[25]   Deng L., Seltzer M., Yu D. 'Binary coding of speech spectrograms using a deep auto-encoder' in Kobayashi T., Hirose K., Nakamura S. (eds.). *INTERSPEECH 2010, 11th Annual Conference of the International Speech Communication Association.* September 26-30, Makuhari, Japan: ISCA Archiv; 2010.

[26]   Bao Y., Jiang H., Dai L., *et al.* 'Incoherent training of deep neural networks to de-correlate bottleneck features for speech recognition'. 2013 IEEE International Conference on Acoustics, Speech and Signal Processing; 2013. pp. 6980–4.

[27]   Garcia-Romero D., Zhang X., McCree A. 'Improving speaker recognition performance in the domain adaptation challenge using deep neural networks'. *2014 IEEE Spoken Language Technology Workshop*; 2014. pp. 378–83.

[28]   Lei Y., Scheffer N., Ferrer L. 'A novel scheme for speaker recognition using a phonetically-aware deep neural network in ICASSP'. *2014 IEEE International Conference on Acoustics, Speech and Signal Processing*; 2014. pp. 1695–9.

[29]   McLaren M., Lei Y., Ferrer L. 'Advances in deep neural network approaches to speaker recognition'. *2015 IEEE International Conference on Acoustics, Speech and Signal Processing*; 2015. pp. 4814–8.

[30]   Garcia-Romero D., McCree A. 'Insights into deep neural networks for speaker recognition'. INTERSPEECH 2015, 16th Annual Conference of the International Speech Communication Association, Dresden; Germany; 2015. pp. 6–10.

[31]   Yaman S., Pelecanos J., Sarikaya R. 'Bottleneck features for speaker recognition'. Proceedings of Odyssey 2012. International Speech Communication Association; 2012.

[32]   Lozano-Diez A., Silnova A., Matějka P. 'Analysis and optimization of bottleneck features for speaker recognition'. Proceedings of Odyssey 2016. International Speech Communication Association; 2016.

[33]   Matějka P., Zhang L., Ng T. 'Neural network bottleneck features for language identification'. Proceedings of Odyssey 2014. vol. 2014. International Speech Communication Association; 2014. pp. 299–304.

[34]   Variani E., Lei X., McDermott E. 'Deep neural networks for small footprint text-dependent SPEAKER verification'. Proceedings of ICASSP; 2014.

[35]   Heigold G., Moreno I., Bengio S. 'End-to-end text-dependent SPEAKER verification'. Proceedings of ICASSP; 2016.

[36]   Snyder D., Garcia-Romero D., Povey D. 'Deep neural network Embeddings for Text-Independent SPEAKER verification'. Proceedings of Interspeech 2017; 2017.

[37]   Prince S.J.D. 'Probabilistic linear discriminant analysis for inferences about identity'. Proceedings of International Conference on Computer Vision (ICCV); 2007.

[38]   Snyder D., Ghahremani P., Povey D. 'Deep neural network-based speaker embeddings for end-to-end speaker verification'. 2016 IEEE Spoken Language Technology Workshop; 2016. pp. 165–70.

[39]   Rohdin J., Silnova A., Diez M., *et al. End-to-end DNN based speaker recognition inspired by i-vector and PLDA*. ArXiv e-prints [online]. 2017. Available from https://arxiv.org/abs/1710.02369.

[40]   Yegnanarayana B., Prasanna S.R.M., Zachariah J.M., Gupta C.S. 'Combining evidence from source, suprasegmental and spectral features for a fixed-text speaker verification system'. *IEEE Transactions on Speech and Audio Processing*. 2005;13(4):575–82.

[41]   Furui S. 'Cepstral analysis technique for automatic speaker verification'. *IEEE Transactions on Acoustics, Speech, and Signal Processing*. 1981;29(2):254–72.

[42]   Ramasubramanian V. 'Text-dependent speaker-recognition systems based on one-pass dynamic programming algorithm'. Proceedings of the IEEE

International Conference on Acoustics, Speech and Signal Processing; 2006. pp. 901–4.

[43]   Rabiner L.R. 'A tutorial on hidden Markov models and selected applications in speech recognition'. *Proceedings of the IEEE*. 1989;77(2):257–86.

[44]   Project C. *CAVE – The European Caller Verification Project* [online]. Available from http://www.ptt-telecom.nl/cave/.

[45]   Campbell J.P. 'Testing with the YOHO CD-ROM voice verification corpus'. Proceedings of the IEEE International Conference on Acoustics, Speech, and Signal Processing; 1995. pp. 341–4.

[46]   Leonard R.G., Doddington G. *TIDIGITS (LDC93S10)* [online]. Available from http://www.ldc.upenn.edu.

[47]   Garofolo J.S., Lamel L.F., Fisher W.M., *et al. TIMIT acoustic-phonetic continuous speech corpus (LDC93S1)* [online]. Available from http://www.ldc. upenn.edu.

[48]   Campbell J., Higgins A. *YOHO speaker verification (LDC94S16)* [online]. Available from http://www.ldc.upenn.edu.

[49]   Woo R., Park A., Hazen T.J. 'The MIT mobile device speaker verification corpus: data collection and preliminary experiments'. Proceedings of IEEE Odyssey; 2006.

[50]   Fierrez J., Ortega-Garcia J., Torre Toledano D., Gonzalez-Rodriguez J. 'Biosec baseline corpus: a multimodal biometric database'. *Pattern Recognition*. 2007;40(4):1389–92.

[51]   Khoury E., Vesnicer B., Franco-Pedroso J. 'The 2013 speaker recognition evaluation in mobile environment'. 2013 International Conference on Biometrics ICB; IEEE; 2013. pp. 1–8.

[52]   Lee K.A., Larcher A., Wang G., *et al.* 'The RedDots data collection for speaker recognition'. Sixteenth Annual Conference of the International Speech Communication Association; 2015.

[53]   Larcher A., Lee K.A., Ma B., *et al.* 'RSR2015: database for text-dependent speaker verification using multiple pass-phrases'. Thirteenth Annual Conference of the International Speech Communication Association; 2012.

[54]   Larcher A., Lee K.A., Ma B., Li H. 'Text-dependent speaker verification: classifiers, databases and RSR2015'. *Speech Communication*. 2014;60:56–77.

[55]   Toledano D.T., Fernández Pozo R., Hernández Trapote Álvaro., Hernández Gómez L. 'Usability evaluation of multi-modal biometric verification systems'. *Interacting with Computers*. 2006;18(5):1101–22.

[56]   Zeinali H., Burget L., Cernocky J. 'A multi purpose and large scale speech corpus in Persian and English for speaker and speech recognition: the DeepMine database'. Proc. ASRU 2019: The 2019 IEEE Automatic Speech Recognition and Understanding Workshop; 2019.

[57]   Zeinali H., Lee K.A., Alam J., Burget L. 'Short-duration speaker verification (SdSV) challenge 2020: the challenge evaluation plan. arXiv preprint arXiv:1912.06311'. 2020.

[58]   Gonzalez-Rodriguez J., Rose P., Ramos D., Toledano D.T., Ortega-Garcia J. 'Emulating DNA: rigorous quantification of evidential weight in transparent

and testable forensic speaker recognition'. *IEEE Transactions on Audio, Speech and Language Processing*. 2007;15(7):2104–15.

[59]   Cook R., Evett I.W., Jackson G., Jones P.J., Lambert J.A. 'A model for case assessment and interpretation'. *Science & Justice*. 1998;38(3):151–6.

[60]   Cook R., Evett I.W., Jackson G., Jones P.J., Lambert J.A. 'A hierarchy of propositions: deciding which level to address in casework'. *Science & Justice*. 1998;38(4):231–9.

[61]   Saunders C.P., Ommen D.M. 'Building a unified statistical framework for the forensic identification of source problems'. *Law Probability and Risk*. 2018;17(2):179–97.

[62]   Willis S. ENFSI guideline for the formulation of Evaluative reports in forensic science. monopoly project MP2010: the development and implementation of an ENFSI standard for reporting evaluative forensic evidence. European Network of Forensic Science Institutes; 2015.

[63]   Doddington G., Liggett W., Martin A. Sheeps, goats, lambs and wolves: a statistical analysis of SPEAKER performance in the NIST 1998 SPEAKER recognition evaluation. Proc. of ICSLP; 1998.

[64]   Champod C., Evett I.W., Jackson G. 'Establishing the most appropriate databases for addressing source level propositions'. *Science & Justice*. 2004;44(3):153–64.

[65]   Li P., Fu Y., Mohammed U., *et al.* 'Probabilistic models for inference about identity'. *IEEE transactions on pattern analysis and machine intelligence*. 2010;34(1):144–57.

[66]   Mclaren M., Lawson A., Ferrer L. 'Trial-based calibration for speaker recognition in unseen conditions'. Odyssey; 2014.

[67]   Ferrer L., Nandwana M.K., McLaren M., Castan D., Lawson A. 'Toward fail-safe speaker recognition: trial-based calibration with a reject option'. *IEEE/ACM Transactions on Audio, Speech, and Language Processing*. 2018;27(1):140–53.

[68]   Meuwly D. Reconaissance de Locuteurs en Sciences Forensiques: L'apport d'une Approache Automatique.[Ph.D. thesis]. IPSC-Universite de Lausanne; 2001.

[69]   Navratil J., Ramaswamy G. 'The awe and mystery of t-norm'. *Proceedings of the European Conference on Speech Communication and Technology*; 2003. pp. 2009–12.

[70]   Gonzalez-Rodriguez J., Fierrez-Aguilar J., Ramos-Castro D., Ortega-Garcia J. 'Bayesian analysis of fingerprint, face and signature evidences with automatic biometric systems'. *Forensic Science International*. 2005;155(2–3):126–40.

[71]   Egli N. *Interpretation of Partial Fingermarks Using an Automated Fingerprint Identification System. Institute de Police Scientifique, Ecole de Sciences Criminelles*; 2009.

[72]   Haraksim R., Ramos D., Meuwly D., Berger C.E.H. 'Measuring coherence of computer-assisted likelihood ratio methods'. *Forensic Science International*. 2015;249:123–32.

[73]   Ramos D., Maroñas-Molano J., Lozano-Diez A. 'Subsidia: Tools and re-
        sources for speech sciencies' in Lahoz-Bengochea J.M., Ramón R.P. (eds.).
        *Bayesian Strategies for Likelihood Ratio Computation in Forensic Voice
        Comparison with Automatic Systems Subsidia: Tools and Resources fot the
        Speech Sciences.* Universidad de Málaga; 2017. pp. 89–95. Available from
        https://riuma.uma.es/xmlui/handle/10630/18177.

[74]   Ramos-Castro D., Gonzalez-Rodriguez J., Montero-Asenjo A. 'Suspect-
        adapted MAP estimation of within-source distributions in generative like-
        lihood ratio estimation'. *Proc. of IEEE Odyssey 2006, the Speaker and
        Language Recognition Workshop*; 2006.

[75]   Brummer N., Garcia-Romero D. 'Generative models for unsupervised score
        calibration'. ICASSP 2014; 2014. pp. 1680–4.

[76]   Brummer N., van Leeuwen D., Swart A. *A comparison of linear and non-
        linear calibrations for speaker recognition.* Odyssey; 2014. pp. 14–18.
        Available from https://www.isca-speech.org/archive/odyssey_2014/index.
        html.

[77]   Pigeon S., Druyts P., Verlinde P. 'Applying logistic regression to the fu-
        sion of the NIST'99 1-Speaker submissions'. *Digital Signal Processing.*
        2000;10(1):237–48.

[78]   Brümmer N., Burget L., Cernocky J., *et al.* 'Fusion of heterogeneous speaker
        recognition systems in the STBU submission for the NIST speaker recogni-
        tion evaluation 2006'. *IEEE Transactions on Audio, Speech, and Language
        Processing.* 2007;15(7):2072–84.

[79]   Brümmer N., du Preez J. 'Application-independent evaluation of speaker de-
        tection'. *Computer Speech & Language.* 2006;20(2–3):230–75.

[80]   Gonzalez-Rodriguez J., Rose P., Ramos D., Toledano D.T., Ortega-Garcia J.
        'Emulating DNA: rigorous quantification of evidential weight in transpar-
        ent and testable forensic speaker recognition'. *IEEE Transactions on Audio,
        Speech and Language Processing.* 2007;15(7):2104–15.

[81]   Morrison G.S. 'Tutorial on logistic-regression calibration and
        fusion:converting a score to a likelihood ratio'. *Australian Journal of Forensic
        Sciences.* 2013;45(2):173–97.

[82]   van Leeuwen D., Brümmer N. 'An introduction to application-independent
        evaluation of speaker recognition systems' in Müller C. (ed.). *Speaker
        Classification. vol. 4343 of Lecture Notes in Computer Science/Artificial
        Intelligence.* Heidelberg - Berlin - New York: Springer; 2007.

[83]   Ramos D., Gonzalez-Rodriguez J., Zadora G., Aitken C. 'Information-
        theoretical assessment of the performance of likelihood ratio computation
        methods'. *Journal of Forensic Sciences.* 2013;58(6):1503–18.

[84]   Vergeer P., Bolck A., Peschier L.J.C., Berger C.E.H., Hendrikse J.N.
        'Likelihood ratio methods for forensic comparison of evaporated gasoline
        residues'. *Science & Justice.* 2014;54(6):401–11.

[85]   Morrison G.S. 'Likelihood-ratio-based forensic speaker comparison using
        parametric representations of vowel formant trajectories'. *The Journal of the
        Acoustical Society of America.* 2009;125(4):2387–97.

[86] Zadora G., Martyna A., Ramos D. *Statistical Analysis in Forensic Science: Evidential Values of Multivariate Physicochemical Data*. John Wiley and sons; 2014.

[87] Ramos D., Franco-Pedroso J., Lozano-Diez A., Gonzalez-Rodriguez J. 'Deconstructing Cross-Entropy for probabilistic binary classifiers'. *Entropy*. 2018;20(3):208.

[88] Duda R.O., Hart P.E., Stork D.G. *Pattern classification*. Wiley; 2001.

[89] Ramos D., Gonzalez-Rodriguez J. 'Reliable support: measuring calibration of likelihood ratios'. *Forensic Science International*. 2013;230(1–3):156–69.

[90] Nautsch A., Meuwly D., Ramos D., Lindh J., Busch C. 'Making likelihood ratios digestible for cross-application performance assessment'. *IEEE Signal Processing Letters*. 2017;24(10):1552–6.

Chapter 3

Voice biometrics: attacker's perspective

Priyanka Gupta[1] and Hemant A. Patil[1]

Abstract

Voice biometric systems, also known as automatic speaker verification (ASV) systems, adopt specialized strategies to authenticate enrolled speakers by means of their claimed identities. In this context, many countermeasures against various spoofing attacks have been proposed during three recent challenge campaigns, namely ASVspoof 2015, ASVspoof 2017, and ASVspoof 2019. Nevertheless, boosting the resiliency of ASV systems just by focusing on the development of countermeasures for anti-spoofing is not enough. To that effect, considering the attackers' perspectives has become definitively crucial. In particular, there have been numerous possibilities to attack ASV systems, and hence, their security can be boosted if the attackers' perspectives are taken into account beforehand, uncovering possible vulnerabilities and loopholes. Thus, this chapter intends to provide insights and understanding into the attackers' possible perspectives, potentially helping in the identification of hidden ASV systems' weaknesses. We present details on different attacks based on the extent of attackers' accessibility to the ASV systems, considering direct and indirect attempts. Apart from the attacks, the threats due to unprotected speech corpora are also discussed. Unprotected speech corpora and ASV systems enable to search for information about a speaker on the Internet, and in this context, privacy-preserving techniques can prevent attackers from getting enrolled speakers' information. Consequently, this chapter additionally discusses various technological challenges occasionally faced by attackers, allowing for their positive exploration to come up with better defense mechanisms.

Keywords: Voice biometrics, anti-spoofing, attackers' perspectives, direct vs. indirect attacks, target selection, adversarial attacks.

[1]Speech Research Lab Dhirubhai Ambani Institute of Information and Communication Technology (DAIICT), Gandhinagar, India

3.1   Introduction

Biometric systems can be based on fingerprints, iris, DNA, or speech, where speech is the most intuitive, simple, and easy to produce characteristic. Humans and animals use speech and vocalization, respectively, to distinguish each other. Voice-based biometric systems, also known as automatic speaker verification (ASV) systems, deal with verifying speakers' claimed identities with the help of machines [1]. In this chapter, both the terms have been used interchangeably. Since ASV systems have been used for banking transactions and access to buildings associated with classified information, for instance, only authorized legitimate or *genuine* users are granted access. Nevertheless, some *impostors*, other than zero-effort impostors, deliberately try to fool the ASV system in order to gain an unauthorized access. The deliberate attempts made by the impostor, playing the role of an attacker, are called *attacks*. Besides, due to the recent commercial success of several intelligent personal assistants (IPAs), also known as voice assistants, such as SIRI, Amazon Alexa, Google Home, and so on, many voice-enabled devices in Internet of Things have also been prone to spoofing attacks [2]. Our aim of introspecting the attackers' perspective is to be able to identify the security vulnerabilities of current ASV systems, consequently designing robust countermeasures (CMs). This is crucial nowadays since those systems are based on neural network approaches and can be successfully attacked by using a number of approaches, including the strategies presented in this chapter. If an attack succeeds in spoofing the ASV system, better CMs would be integrated for real-time use. In addition, the same type of attack can be used as a reference testing point for the further design of ASV anti-spoofing systems.

ASV systems do suffer from vulnerabilities to attacks, and therefore, they can be compromised at various stages. The attacks may be categorized as direct and indirect attacks, depending on the extent of the attacker's accessibility to the ASV framework, as shown in Figure 3.1 [3]. In the figure, the sub-modules, such as the

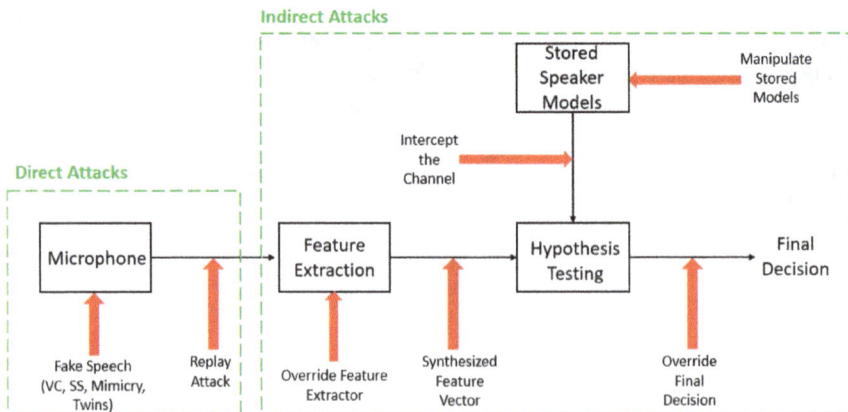

Figure 3.1   Direct vs. *indirect attacks. After [3]*

Direct/Black Box Attacks | Indirect Attacks

Adversarial
Attacks

Grey Box Attacks

Hardware
Attacks

Hardware
Attacks

Adversarial
Attacks

White Box Attacks

Spoofing

Attack on Corpus

- No access to the
  ASV system.
- Can use a similar
  ASV to perform
  'target selection',
  before generating
  adversarial
  examples.

- Twins
- Professional
  mimics
- VC
- SS
- Replay

- Side-channel
  attacks: Timing,
  power, cache trace.
- Manipulated
  microphone.

- Fault injection
  to misclassify
  or corrupt the
  ASV system.

- Can be used to generate
  improved and targeted
  adversarial examples.
- Personal information of speakers
  can be extracted.

- Knows the target
  model.
- Has access to
  classifier, gradient
  and training data.

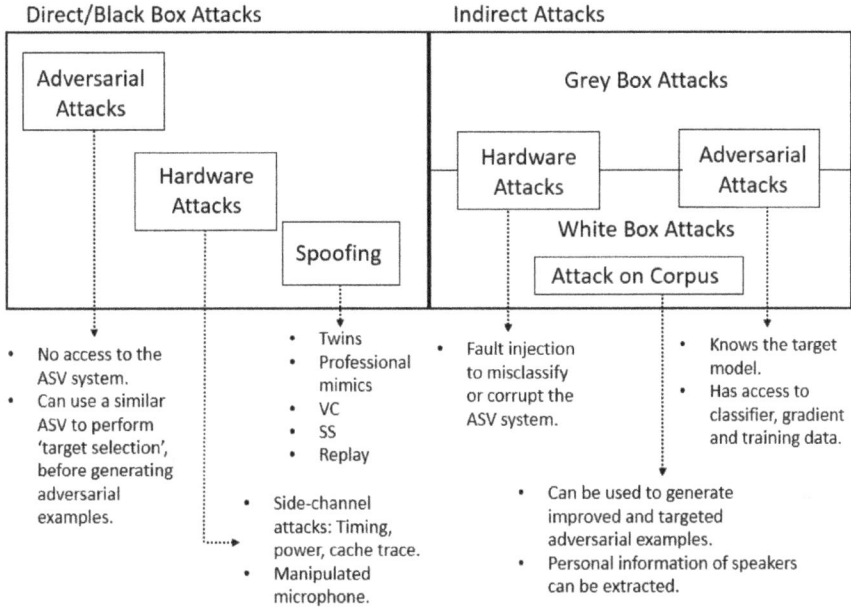

Figure 3.2   Classification of various attacks on an ASV system

feature extractor, can be manipulated to override the attacker's malicious features. Apart from manipulating the sub-modules, the communication between them can also be intercepted to the attacker's benefit.

The various types of attacks have been classified, as shown in Figure 3.2. Direct attacks do not require access to the internal sub-modules of the target ASV system to be granted, thus they are also called black box attacks. Contrary to this, indirect attacks are possible whenever the attacker has partial or complete knowledge of the target ASV system. The various types of attacks are discussed in the upcoming subsections of this chapter.

The attacker's perspective can be learnt if the attacker's intentions are also kept in mind. Knowing the specific attacker's purpose can help us to narrow down the number of possible attacks. Depending on the motive, the attacker will craft a particular attack which fits best to achieve the goal, which can be classified into three categories, as shown in Figure 3.3. The first goal, allowing for an authorization to be provided in benefit of an unauthorized user, aims to affect the equal error rate (EER) by increasing the false acceptance rate (FAR).

The second purpose, which prevents authorization for a genuine user, affects the EER by increasing the false rejection rate (FRR). The third one, aiming to gain information about genuine speakers by using their information illegally, can be done by selecting the most vulnerable genuine speaker from the training dataset.

This chapter is a study about various attacking techniques and approaches. To that effect, the attacks have been classified as direct and indirect. With that in mind,

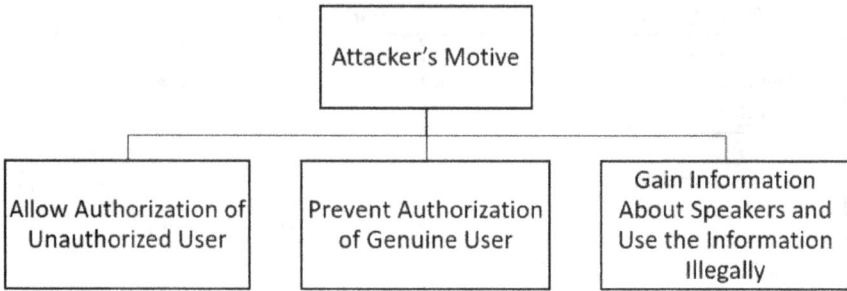

Figure 3.3   Motivation for attacks on ASV

the remainder of this chapter is organized as follows. Section 3.2 describes direct attacks in detail, including all kinds of spoofing attacks, black box hardware attacks, and adversarial attacks. Section 3.3 presents indirect attacks, including attacks on corpus by using the target selection technique which helps the attackers to optimize their actions toward the most vulnerable speaker. Section 3.3 discusses gray box hardware attacks followed by gray box and black box adversarial attacks. A detailed discussion on various technological challenges which could be faced by the attacker are presented in Section 3.4. The chapter closes with Section 3.5 describing the conclusions and possible future works.

3.2   Direct attacks

Direct attacks are events implemented and executed without using knowledge from the internal ASV system architecture. Therefore, in the event of a direct attack, the attacker does not actually break through, or fool, any internal subsystem in the target ASV system. Instead, microphone and transmission-level spoofing attacks are performed. To that effect, no previous knowledge of the ASV system in question is necessary to successfully attack it. This is the reason why such an event is also known as *black box attack*. Thus, this kind of attack poses a significant threat to the security of the ASV system due to their ease of execution. Types of direct attacks are spoofing attacks, hardware attacks, and adversarial attacks, as shown in Figure 3.2.

3.2.1   Spoofing attacks

Spoofing attacks are also called presentation attacks. Particularly, voice conversion (VC) [4, 5], speech synthesis (SS) [6, 7], replay [8–10], twins [3], and impersonation [11] are predominant examples of spoofing attacks on ASV systems [12]. We will discuss some of the spoofing attacks in this section.

Impersonation: Modifying one's voice to sound like a target speaker's voice consists of an impersonation in which the attacker mimics the characteristic of the speaker just by using personal skills, without any special technology. It can be performed by professional mimicry artists or, even more so, by identical twins, who exploit behavioral or physiological [11] characteristics, respectively, to sound like the target speaker. Imitation of speech includes adapting high-level features, such as the prosody, accent, and pronunciation. Such imitation may mislead human perception; however, it is less effective in attacking ASV systems because most of them receives inputs from short-term spectral features to make decisions. Regardless of this fact, we still consider impersonation as a threat to ASV systems, because spectral features are found to be similar for an identical biological twin-pair [13]. Notably, characteristics such as pitch (F_0) contour, formant contour, and spectrograms are very similar for identical twin-pairs [14]. A real case example of twins fraud occurred in HSBC Bank, where a BBC journalist and his non-identical twin spoofed the bank's voice authentication system [15, 16].

Speech synthesis (SS): Also known as text-to-speech synthesis, this kind of attack uses text as input to generate speech as output. It mirrors a human speech production mechanism system, i.e., vocal tract and glottal information, to pose a threat to the ASV system. Due to technological advances, the obtained speech quality sounds considerably natural. Some of the advances which enable this threat are unit selection [17], statistical parametric [17], hybrid [18], and deep neural network (DNN)-based methods. Recently, deep learning-based techniques such as generative adversarial networks (GANs) [19] have been able to perform very well in terms of fooling ASV systems.

Voice conversion (VC): This attack aims to convert a source speaker's speech to sound like a target speaker's speech [20, 21]. Signal processing techniques such as vector quantization [22] and frequency warping [23] have been used to achieve successful VC strategies. DNN-based VC utilizing wavenets and GANs have received significant attention from the research community.

Replay: It is one of the most convenient attacks to execute but difficult to detect. The attacker uses a pre-recorded speech from the target speaker to get access through the ASV system. Compared with the recorded speech, the genuine data differs only slightly. The differences are due to the impulse response of the recording device and the recording environment. Replay attacks have been a great threat due to advent of high-quality recording devices and the careful choice of the recording environment in order to minimize acoustical noise [24–26].

Keeping in mind the vulnerability of ASV systems to so many spoofing attacks, ASVspoof challenges were organized by INTERSPEECH chairs to boost related research. With ASVspoof 2015 challenge, various CMs were proposed based on different forms of feature extraction techniques on a standard dataset [27–33]. Most of the participant teams centered on signal processing-based research proposals, to design feature sets, and Gaussian mixture models, as pattern classifiers for the genuine *vs.* spoofed speech detection (SSD) problem. In addition, multiple CMs for the replay spoof detection were proposed in the ASVspoof 2017 competition [34–36]. In that challenge campaign, there was a paradigm shift from signal processing-based work to sophisticated deep learning approaches. In ASVspoof 2015 and ASVspoof 2017 challenges, the assessment of CMs was performed independently from ASV systems using EER curves. SSD systems were used before the ASV system to detect spoofed speech using CMs, thus making it a two-class problem, as shown

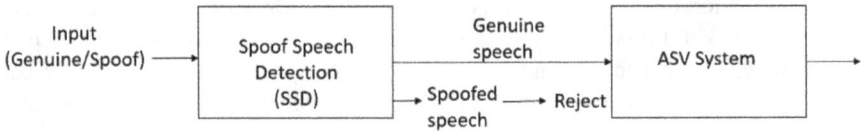

Figure 3.4   SSD system for ASV system

in Figure 3.4. However, this type of assessment is not suited for real-world applications involving user authentication. Therefore, in ASVspoof 2019 challenge, an ASVspoof SSD-centric assessment called *tandem detection cost function* (t-DCF) was proposed to improve the overall reliability of ASV systems, where the SSD system is assessed in tandem with the ASV system.

Significant recent initiatives to exploit state-of-the-art approaches exist in the ASV literature [37], for instance, National Institute of Standards and Technology (NIST) Speaker Recognition Evaluation (SRE) 2019 [38], first Voice Privacy Challenge 2020 [39], Short-Duration Speaker Verification Challenge [40], and Far-Field Speaker Verification Challenge [41]. Knowing the prospect of an attacker is important to have more protected ASV systems with the existing spoofing CMs. Furthermore, the ISO/IEC 30107-1 standard was laid specifically to provide a foundation for spoofing attacks on the data capture subsystem, i.e., at the microphone level of the ASV system, also called presentation attack detection.

Out of the two types of attacks, i.e., direct and indirect, the former threats ASV systems majorly. This is because spoofing attacks are performed at the microphone and transmission levels of the ASV system and, hence, no prior knowledge of the ASV system in question is needed to attack it successfully. Technologies such as VC and SS were not designed originally for the purpose of attacking an ASV system, and their ability to spoof the ASV system is stated as a coincidental side product [42]. Moreover, those techniques simply aim to increase the system's FAR. Therefore, such type of spoofing attacks are called *non-proactive* [42]. However, it is crucial to state that although these methods were originally developed for different applications, with no connection to attacks on ASV systems, an attacker with a malicious intent cannot be made to refrain from using any of those methods. Moreover, due to their ease of execution, these types of attacks, specially replay ones, are more convenient for the attacker. Security requirement analysis in an ASV system can be performed whenever various possible approaches and attacker's perspectives are known beforehand. Therefore, possible vulnerability aspects should be examined in order to make an ASV system robust against spoofing.

3.2.2   Black box hardware attacks

Hardware attacks are a threat to ASV systems whenever the implementation of the CMs and any security algorithms has hardware flaws. Internal knowledge, such as the type of feature and classifier used, is not useful to execute black box hardware attacks. To that effect, such attacks are independent of any knowledge on the ASV system. Hardware outputs, such as power, timing, and cache traits, can be monitored and observed by the attacker to get enough information about the ASV system in order to attack it. Such attacks, predominant where security is achieved through cryptographic techniques, are called *side-channel attacks*. Simple power analysis (SPA) and differential power analysis (DPA), for instance, are classic examples of such type of attack [43, 44]. Cryptographic methodologies, such as elliptic curve cryptography, are said to be computationally unbreakable till now; however, their vulnerability gets highlighted when it comes to side-channel attacks, such as SPA and DPA [45]. DPA attacks have the properties of signal processing and error correction. Consequently, they can obtain information about secret keys from implementations containing too much noise, by using SPA. Hardware attacks are difficult to execute. They require the attacker to be an expert in the field of hardware. Technological challenges for the attacker increase whenever cryptography implementations are used to achieve the desired security. Furthermore, apart from side-channel attacks, the inputs to the sensor device, such as the microphone in the case of ASV systems, to the biometric can also be tracked, or worse, manipulated.

3.2.3 Black box adversarial attacks

Adversarial attacks on ASV systems are executed by synthesizing *artificial signals* which sound perceptually natural to the human ear. This artificially generated signals are called *adversarial examples*. Their aim is to make the classifiers misclassify speech data, thereby providing incorrect decisions at their output [46]. Even a well-trained neural network model can be vulnerable to adversarial attacks [47]. In [48], the authors prove the weakness of automatic speech recognition neural network model under adversarial attacks, where an adversarial example can be synthesized as any sentence.

Studying black box scenarios is crucial because once a successful attack is found, it ensures the success of a white box attack to a great extent, since black box attack requires less information and is more difficult to perform. Adversarial attacks in black box scenario are performed without access to the internals of the target model [49]. With just pairwise input–output information from training and testing data, a similar substitute model can be trained. The substitute model is then used to craft adversarial examples to attack the target system. This approach to use a similar ASV to attack a target ASV system can be used to perform many other types of attacks. An adversarial black box attack, named "FakeBob," is analyzed in [50] under various aspects such as transferability of the attack, imperceptibility by the human ear, and over-the-air practicality. Another example of a black box adversarial attack is demonstrated by feedback-controlled VC framework in [51]. Here, we will discuss about *black box adversarial attack by substitute model training* [52].

Black box adversarial attack by substitute model training: In this approach, the DNN-based target ASV is termed the "oracle O." The adversary, i.e., attacker, has only one power in the black box attack scenario: the attacker can access the label $\bar{O}(x)$ assigned for an input signal x. Now, the attacker aims to train a substitute model without even a real labeled dataset. This is performed by first generating synthetic inputs and observing their corresponding labels from the target ASV, i.e., oracle. Using this synthetic dataset, the attacker designs an approximate substitute model F of O. After the substitute model F has been created, the next required step is to generate adversarial examples which should be misclassified by F. Now, since the attacker has full knowledge of the substitute ASV F, the attacker can craft adversarial examples which would be misclassified by F. As long as the *transferability property* holds between F and O, adversarial samples crafted for F will also be misclassified by O. Furthermore, in [53], two types of black box attack scenarios were considered: a cross-dataset attack and a cross-feature set attack.

In the case of *cross-dataset attacks*, the adversarial examples are generated by using a model trained on dataset A, and then the generated adversarial examples are used to attack a model trained on dataset B. For example, in [53], two models were trained on two different datasets. Model A was trained on the YOHO corpus and model B was trained on the NTIMIT corpus. The features used for training model A and model B were Mel frequency cepstral coefficient (MFCC) and Mel-spectrum features, respectively. A value of accuracy of 85% was achieved when attack was implemented on Model A using clean NTIMIT test dataset. However, the accuracy dropped to 58.93% when the attack on Model A was performed using the crafted adversarial examples on NTIMIT test dataset.

In the case of *cross-feature attacks*, adversarial examples are created by using one set of acoustic features, and then, a model that was trained on a different set of acoustic features is attacked. For example, in [53], two models were trained on the same YOHO dataset. However, one model was trained using Mel-spectrum features and the other model was trained using MFCC. An accuracy of 81% was observed when the MFCC-based model was attacked using clean YOHO test dataset. However, the accuracy dropped to 62.25% when crafted adversarial examples based on YOHO dataset were used in the attack.

Generating adversarial examples: To generate adversarial examples, various approaches have been proposed. Here, we describe the *fast gradient sign method (FGSM)* to craft adversarial examples. An adversarial example w.r.t. an original speech signal x can be represented as

$$\bar{x} = x + \delta \tag{3.1}$$

where δ is so small that $\bar{x}$ is perceptually identical to x. However, δ is large enough to cause misclassification. The optimum value of δ can be computed by solving the optimization problem [54]

$$\max_{\delta \in \Delta} \text{Loss}(\theta, x + \delta, y_x) \tag{3.2}$$

where Loss(.) is the loss function, θ is the model parameter vector, Δ is the set of allowed perturbations, and y_x is the label of x. We discuss two of the popular methods to solve the optimization problem shown in (3.2). The original signal x is manipulated by adding or subtracting a small error ϵ.

The FGSM is one of the simplest methods to generate adversarial examples. The sign of the gradient ∇ determines whether we add or subtract the error ϵ. Adding errors in the direction of the gradient means that the signal is intentionally altered so that the model classification fails. The following equation describes FGSM [54]:

$$\delta = \epsilon \cdot \text{sign}(\nabla_x \text{Loss}(\theta, x, y_x)) \tag{3.3}$$

where ϵ is the small error induced deliberately, and ∇_x is the gradient of the cost function w.r.t. x. If the increase in loss is caused by manipulating x as *increase* to x, the sign of the gradient is positive, i.e., $+1$. Contrary to this, if the increase in loss is caused by manipulating x as *decrease* to x, then the sign of the gradient is negative, i.e., -1. Therefore, the adversarial spectral feature is given by

$$\bar{x} = x + \epsilon \cdot \text{sign}(\nabla_x \text{Loss}(\theta, x, y_x)) \tag{3.4}$$

The vulnerability occurs when the relationship between input signal and class score is treated as *linear*. Neural network architectures which encourage linearity, such as long short-term memory (LSTM), maxout networks, networks with ReLU activation units, or other linear machine learning algorithms, such as logistic regression, are vulnerable to the gradient sign method [55]. The models in these architectures can be deceived by moving the $\bar{x}$ into areas outside the data distribution, as *outliers*. Furthermore, those adversarial examples do not turn out to be specific only to a given neural network architecture. Instead, they can be reused to deceive neural networks with a different architecture. Another method to craft adversarial examples is the projected gradient descent (PGD) method.

PGD method: The PGD method is an iterative method in which the original input ($x = x_0$) is iteratively updated as follows:

$$x_{k+1} = \text{clip}(x_k + \alpha \cdot \text{sign}(\nabla_{x_k} \text{Loss}(\theta, x_k, y_{x_k}))), \quad k = 0, ..., K - 1 \tag{3.5}$$

where α is the step size, K is the number of iterations, and the clip$(\cdot)$ function applies element-wise clipping such that $\|x_k - x\|_\infty < \epsilon$, $\epsilon \leq 0 \in R$, being x_k the final perturbed spectral feature. The PGD method is nothing but an iterative form of small-step FGSM, where the perturbed input is forced to stay within the admissible set ∇ at every step. It allows for more effective attacks, but it is more computationally expensive than the FGSM due to its iterative nature. The PGD method is bound to be limited to stick at local optima of the loss function. Therefore, random restarts are incorporated, which makes the PGD to be executed a certain number of times in one run [56].

Additionally, in paper [57], trained CM models were attacked by using adversarial attacks with FGSM and PGD methods. To make the level of manipulative power of the PGD attack consistent with the FGSM attack, the step-size α and the parameter ϵ in the PGD attack scenario are made to satisfy the relationship ϵ = number of iterations $* \alpha$. The performance of the black box attack was observed in three settings of CM models: LCNN-big, LCNN-small, and SENet12. The attack settings were LCNN-big with LCNN-small, LCNN-big with SENet12, and LCNN-small with SENet12. On the basis of EER, the PGD method generated better adversarial examples than the FGSM method. Furthermore, LCNN-big presented better results in generating powerful adversarial examples which can further attack smaller models like LCNN-small and SENet12. Therefore, adversarial examples from small models are outperformed by large ones in terms of the attack performance for both the FGSM and PGD methods. Moreover, the authors observed that adversarial attacks were realized much easier under similar model structures.

3.3   Indirect attacks

Indirect attacks are those occurring in system-levels, being feasible whenever the attacker has access to the internal subsystems of the target ASV system. If the attacker has complete knowledge and access to all the subsystems, the attack is termed as a *white box* attack. It represents an ideal scenario for attackers, which is not practically realistic. However, despite their unrealistic nature, these attacks should not be ignored since they represent the worst-case possibility for the security of ASV systems. The robustness of an ASV system should be evaluated against such a worst-case scenario so that the ASV systems, and their associated CMs, are fully prepared to prevent most of the possible attacks.

A more realistic case of indirect attacks is that in which the attacker has partial knowledge of the target ASV system. Such indirect attacks are termed *gray box attacks*. Most of the indirect attacks are gray box attacks due to their realistic nature. An attacker can perform more serious damage to the ASV system security by implementing a gray box attack as compared to a black box attack because more power, i.e., knowledge, on the gray box target ASV system exists. In this section, we will discuss some of the possible indirect attacks.

3.3.1   *Attacks on corpora*

Attacks over unprotected corpora can be used to determine personal information about speakers. The ISO/IEC International Standard 24745 on Biometric Information Protection [58] enforces that, for full privacy protection, biometric references should be irreversible and *unlinkable*. Unlinkability is achieved if the enrolled users' identity is "un-linked," or hidden, from the linguistic content [59–61]. An unprotected speech corpus, i.e., the biometric reference, and an ASV system enable to search for information about a speaker on the Internet [11, 62]. The study reported in paper [63] deals with matching users' speech to celebrities' speech data on YouTube. Thus, due to the availability of public-domain target speaker data collected from YouTube, termed "found data,", users can find out which celebrity's voice resembles the most to the user's voice. To that effect, an attacker effectively searches for and selects a speaker whose voice mostly matches the attacker's voice. This approach, known as *target selection*, is described ahead.

Target selection: From an attacker's point of view, a *target* is the most vulnerable speaker. Target selection is an approach which can be used by the attacker to identify the most vulnerable speaker to be subjected to spoofing attacks. Choosing the most vulnerable speaker allows the attacker to focus on and optimize the intended attack toward the target speaker only, thereby increasing the FAR. One such example includes enhancing mimicry attack by selecting the most vulnerable target [64]. Moreover, by performing target selection prior to an attack, spoofing attacks against a closed source target ASV system can be improved. To that effect, probability of a successful attack increases whenever the target selection is performed priory.

The target speaker is selected by using a publicly available ASV system, i.e., the attacker's ASV, as shown in Stage 1 of Figure 3.5. The attacker's ASV can be

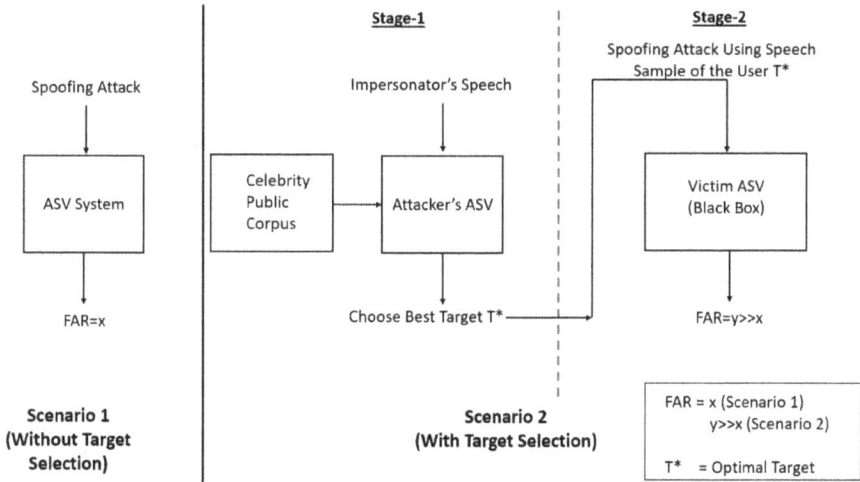

Figure 3.5   Using ASV to attack the victim ASV

trained on the celebrity public corpus [64]. One by one, each of the enrolled speakers' speech data is fed as input to the attacker's ASV system. The aim of the attacker is to impersonate himself/herself as the specified target speaker from the celebrity public corpus. Given $X = [X_1, X_2, ..., X_N]$, as being the set of d-dimensional feature vectors, and H_0 and H_1 as the hypothesis as to whether the speaker is the same and different, respectively, the computation of LLR is given by (3.6) [65]:

$$l = log\frac{p(X \mid H_0)}{p(X \mid H_1)} \tag{3.6}$$

The log-likelihood ratio (LLR) score in (3.1) is compared to a predefined threshold, which then defines the FAR and FRR [3]. The higher the LLR score, the stronger the support for the hypothesis H_0 is, i.e., a higher value of LLR indicates a high probability that the attacker's speech is the same as that of the speaker from the public corpus. To that effect, the LLR measure for choosing a target speaker is computed by using an ASV system, preferably similar to the target ASV system, as shown in Figure 3.5. The speaker with the highest value of LLR score is then chosen as the target speaker.

It is important to note that, even after selecting target speaker in Stage 1 as shown in Figure 3.5, the attacked ASV system might not provide increased FAR, as needed by the attacker. This can happen because of substantial differences between the attacker's ASV and the target ASV. However, this is only a special case which might or might not occur. In general, it is expected that target selection approach, as in Scenario 2 of Section 3.2 and as shown in Figure 3.5, provides a greater value of FAR, than the FAR achieved when no target selection strategy is used.

Depending on their respective LLR scores, and hence their effects on EERs, the pool of speakers in a corpora can be classified as shown in Table 3.1. The speakers in the pool have different vulnerability levels. The most vulnerable speaker, referred to

*Table 3.1   Classification of speakers for target selection to attack ASV system.
After [66]*

Types of speakers in an ASV	Symbolic notation	Vulnerability to ASV
Well-behaved speakers: sheep		Not a vulnerability (low FRR)
Difficult to recognize speakers: goats		Increased FRR
Easy to mimic (easy to attack): lambs		Increased FAR
Successful at imitating other speakers: wolves		Increased FAR

as *lamb* in [66], is the easiest target to imitate from the attacker's perspective. Since the most vulnerable target would have increased FAR, we can say that it is the most accepted speaker by the ASV system. Hence, the speaker with the highest LLR is selected as the target.

The attacker's approach of target selection should not be confused with speaker identification. In the latter, a claimed identity is compared with all the speaker models and the speaker model with the largest closeness to the claimed identity is chosen. In the former, however, there is no single speaker as claimed identity and, hence, the ASV has to be run *iteratively* in order to include all the speakers that are enrolled during the enrollment phase, as shown in Figure 3.5. Moreover, the target speaker chosen is responsible for a maximum FAR, out of all the enrolled speakers in an ASV system.

3.3.2   Gray box hardware attacks

Hardware attacks are predominantly performed on encrypted electronic implementations. Faults are deliberately injected by the attacker to change the electrical characteristic behavior of the circuit used. Only a certain subset of those injection fault attacks are detectable by using side-channel analysis of power consumed and timing parameters. However, in practice, performing side-channel analysis to detect fault injection attacks is cumbersome. An example of fault injection attack on a encrypted hardware is that performed by injecting parametric *Trojan* [67]. With the help of parametric Trojan, the electrical characteristics of the logic gates used in the circuit are altered. This is performed by exploiting doping concentration levels in the active region of the transistors used for that particular target gate. Apart from parametric Trojan, a few more techniques for hardware attacks are shown in Table 3.2.

3.3.3   Gray box and white box adversarial attacks

In gray box and white box adversarial attacks, the attacker has more access to the target ASV as compared to the black box attacks. The perturbation in an adversarial example is small enough but causes the ASV system to make incorrect decisions. Therefore, searching for a suitable δ is the main objective. If the attacker has the ability to perform gray or white box attacks, then the abovementioned challenges can be overcome by the attacker. In papers [69, 70], the adversary is assumed to have access to the physical environment where the actual attack will take place, being capable of computing the room impulse response (RIR) by using a speaker and a microphone. This enables the attacker to launch over-the-air attacks. To keep the perturbation unaffected while being played in the air, the attacker uses the estimated RIR to model the perturbation during generation of adversarial examples.

In paper [53], for white box setting, the authors assumed that the adversary has complete knowledge and control of the network and can access the networks' gradients. Those gradients are then used to generate adversarial examples as input to the ASV system. The adversarial examples are crafted directly on the inputs to the ASV system. System accuracy under white box attack was observed for two datasets: YOHO and NTIMIT. The difference between the accuracies of original test and adversarial test was computed, allowing for an interesting observation: for YOHO dataset with Mel-spectrum features, the difference is 48% and for MFCC features, it is 61.75%. Similarly, for the NTIMIT dataset, the accuracy difference values are 59.86% and 69.96% for Mel-spectrum and MFCC features, respectively. Hence, it can be concluded that the white box attack affects the system accuracy drastically.

For gray box adversarial attacks, the attacker can be assumed to have access to corpus used, as in paper [71]. Using the speaker classification for target selection, as

Table 3.2   Summary of fault injection techniques. Adapted from [68]

Technique	Accuracy (space)	Accuracy (time)	Technical skill	Cost	Hindered by technological advances	Requires knowledge of the implementation	Damage to the device
Underfeeding	High	None	Basic	Low	No	No	No
Clock glitch	Low	High	Moderate	Low	Yes	Yes	No
EM pulses	Low	Moderate	Moderate	Low	No	No	Possibly
Heat	Low	None	Low	Low	Partial	Yes	Possibly
Power supply glitch	Low	Moderate	Moderate	Low	No	Partial	No
Light radiation	Low	Low	Moderate	Low	Yes	No	Yes
Light pulse	Moderate	Moderate	Moderate	Moderate	Yes	Yes	Possibly
Laser beam	High	High	High	High	Yes	Yes	Yes
Focused ion beam	Complete	Complete	Very high	Very high	Yes	Yes	Yes

shown in Figure 3.5, *master voices* (MVs) have been chosen in [71] for dictionary attacks. MVs correspond to the *wolves* in the pool of speakers, i.e., those speakers which can match with a number of other speakers. The results reported in [71] show that adversarial dictionary attacks on an ASV allow for an effective search for MVs. Even in the most conservative setting, it was shown that an MV could match 10% of males and 20% of females. Therefore, dictionary attacks can be seen as valid threat models in the case of gray box adversarial attacks. However, there are two major challenges for the attacker. First, the adversarial examples may suffer from over-the-air problem, wherein the deliberately added perturbation gets distorted. Second, the ASV and the associated CM use signal smoothing so that the effect of perturbation is clipped out.

3.4 Technological challenges

In this section, we discuss various technological challenges that the attacker will have to face in order to attack any ASV system. It means that, if these technical difficulties are resolved or understood well by the attacker, the susceptibility of ASV systems to spoofing attacks will rise significantly. However, it should be noted that such technological challenges can provide future directions to design attack-resilient ASV systems.

3.4.1 Extracting prosodic information

Direct attack by impersonation and mimicry requires professional skillfulness and prosodic information. Due to the limited availability of professional mimicry attackers, synthesizing naturally sounding artificial signals is a challenging task. This is because prosodic information is difficult to incorporate in a single speech frame at the segmental level. In order to capture speech prosody effectively, longer duration of speech (100 ms) at suprasegmental level is required, whereas the state-of-the-art features such as MFCCs and constant Q cepstral coefficients are extracted from 10–30 ms short-time speech frames, i.e., at the segmental level. The hierarchy of various perceptual cues that captures a speaker's identity implicitly is shown in Table 3.3 [72, 73].

3.4.2 Enrolled users with malicious intent

The general assumption is that the attacker is someone who does not belong to the speakers enrolled in the corpus. This assumption rules out the realistic scenario of having an enrolled user with malicious intent. A classic example is of *twins fraud*, where both the co-twins are enrolled speakers in the ASV system. One of the co-twins may impersonate as the other twin. Hence, enrolled co-twins have more power to attack than an external attacker who is not enrolled [78]. The importance of studying twins-based frauds was reported originally in [1]. However, due to unavailability of statistically meaningful twins corpus, this problem has not been given much attention in the ASV literature. The case of HSBC Bank is a real case example of

Table 3.3   Hierarchy of perceptual cues. After [72–77]

Level of difficulty	Perceptual cues	Factors	Algorithmic difficulty
High-level cues (learned traits)	Semantics, dictions, pronunciations, idiosyncrasies	Socio-economic status, education, place of birth	Difficult to extract automatically
	Prosody, rhythm, speed intonation, volume intonation	Personality type, parental influence	
Low-level cues (physical traits)	Acoustic aspect of speech, nasal, deep, breathy, rough	Anatomical structure of vocal apparatus	Easy to extract automatically

twins-based fraud, where a BBC journalist and his non-identical twin spoofed the institutional voice authentication system [15, 16]. The technological challenge in such a case is that, although the malicious twins will be prevented from impersonation by a suitable CM, the CM will also inhibit authentic and zero-effort impostors from being verified and, thus, the FRR will increase. Hence, further study in twins-based fraud in the context of the design of secure voice biometric systems is needed, as originally discussed in [1, 79].

3.4.3   Number of trials permitted on the ASV

For practical situations, a successful ASV system will have an upper limit on the number of trials that a single speaker would be able to perform during verification. This would prevent brute force attacks and *denial of service* attacks, at the same time. The latter attacks are those in which the attacker keeps the ASV system busy with very rapid verification requests. This prevents the genuine users to input their speech sample at all. Hence, if the permitted number of attempts are finite, such kind of attacks cannot be executed.

3.4.4   Minuteness of the perturbation in adversarial attacks

The boon for imperceptibility of the perturbation in adversarial attacks can also become disadvantageous for the attacker. Due to over-the-air effect, the perturbation deliberately induced by the attacker can be suppressed over the transmission channel. Hence, the attack may be unsuccessful, especially in the case of voice assistants or IPAs [80, 81]. Consequently, a critical challenge is faced during black box adversarial attacks. Furthermore, existing CMs use smoothing techniques to smooth out the perturbations in the input signal. Therefore, the perturbation should be such that it bypasses the smoothing techniques [82–85]. In paper [85], adversarial examples are generated by using two methods: FGSM and local distributional smoothness. The ASV is trained by using *adversarial regularization*, which aims to search for a worst spot around the current data point. Furthermore, optimization is performed

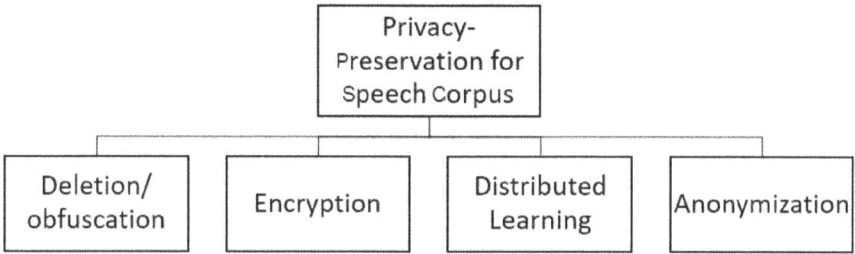

Figure 3.6   Privacy-preservation for speech corpus. After [42]

by using this worst data point. To that effect, the overall model becomes robust to adversarial perturbation.

3.4.5   Privacy preservation of speech and voice privacy

Speech conveys much more information than just the linguistic content. Human beings can perceptually identify speakers by means of their individual voices, even though the voice has no linguistic information pertaining to the person's name or identity. This means that apart from the linguistic content, speech also contains speaker-specific information. Characteristics such as accent, tone, and speaking rate enable the listener to recognize the age, gender, emotional state, health condition, social, and geographic background of a speaker [86, 87]. Therefore, privacy-preservation of speech corpus is crucial. Therefore, if the publicly available corpus is protected in such a way that target selection does not meet the attacker's objectives, then a majority of speaker-specific attacks can be prevented. The protection of the corpus can thus be implemented by using different approaches, as shown in Figure 3.6.

Current methods for privacy preservation of speech corpus can be categorized as deletion, encryption, distributed learning, and anonymization [42], as follows:

- **Deletion/obfuscation:** Deletion methods [88, 89] delete or obfuscate overlapping background noise with a speech signal to the point that the speech becomes completely unintelligible. The approach is inspired by Google's face blurring concept in Street View photographs [88]. The original audio signals are separated into two components: voice and background. The former is "blurred" and then mixed with the latter. This is achieved by using source separation, processing, and remixing. By adopting this method, speaker de-identification is achieved along with content obfuscation. Due to content obfuscation, the signal becomes unintelligible. To that effect, deletion and obfuscation methods are not preferred.
- **Encryption methods:** Encryption methods [90, 91], such as homomorphic encryption [92] and secure multiparty computation [93], enable comparisons and other mathematical computations to be done even on encrypted data, thus alleviating the need for decryption. More importantly, even if the intermediate

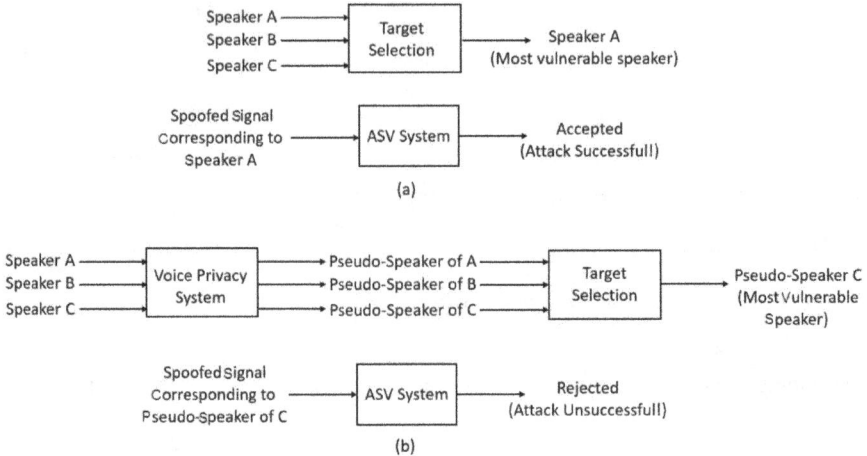

Figure 3.7 *(a) Attack using target selection but without voice privacy system, and (b) attack using target selection with voice privacy system*

encrypted data gets leaked to the attacker, it will not be of any use to the attacker in the encrypted form. However, encryption methods face significant computational complexity, which further requires robust and optimally timed hardware implementations.

- **Distributed learning:** Without having direct access to data, models can be learned from distributed components by using federated learning methods [94]. However, the derived data used for learning, such as the model gradient, can still leak important information [95].
- **Anonymization:** Privacy-preservation of the speech corpus by anonymization defends the ASV system against targeted attacks. In Figure 3.7(a), out of the three types of speakers, i.e., A, B, and C, the most vulnerable is speaker A. In the absence of a voice privacy system, target selection provides correct results according to the vulnerability levels of the actual speakers.

If anonymization is achieved on the corpus by using a voice privacy system, the actual speakers are mapped to their corresponding *pseudo-speakers*. The output of a voice privacy system is a speech utterance which sounds as if it had been spoken by another speaker, known as *pseudo-speaker*. Thus, anonymization alters the identity of the speaker but retains its intelligibility and naturalness. The change in identity modifies the vulnerability levels of the speakers. Therefore, as shown in Figure 3.7(b), after voice privacy is applied, the target selection system yields pseudo-speaker C as the most vulnerable one, whereas in reality, speaker A is the most vulnerable. Hence, the attacker is fooled into believing that the target is pseudo-speaker C. Therefore, the technological challenge faced by the attacker is due to the anonymization provided by the voice privacy system on a speech corpus.

3.5   Conclusions and future work

This chapter discussed different approaches to attack voice biometric systems. Important security gaps in the existing ASV systems have been presented together with detailed comments on the various kinds of possible black, gray, and white box attacks, with emphasis on adversarial attacks. Additional attacks based on inefficiencies in hardware implementation of cryptographic algorithms used for privacy preservation have also been briefly explained. Apart from the security of ASV systems solely, the importance of privacy preservation of speech data and its effect on target selection approach by the attacker were also presented. The chapter concludes with an overview of various technological challenges faced by the attacker. The challenging factors for the attacker can be exploited constructively to build more robust defensive mechanism for ASV systems.

In the future, GANs producing effective adversarial examples can be exploited for improving attacks on ASV systems [54]. GANs generate samples that resemble those produced from the data. They have been extensively used for image processing applications and, recently, in VC, whisper-to-normal speech conversion, and speech enhancement [96–98]. However, their applications to attacks and CMs in ASV systems remain unexplored. Moreover, performance of different types of SSD systems in the presence of the described attacks should be studied and analyzed deeply to allow for replication. There are some attack approaches, such as target selection and black box adversarial attacks using substitute ASV, which are feasible due to the assumption that the attacker can read and obtain all the input–output pairs of the target ASV system. In the case of such a type of approach, the effect of constraining the number of permitted trials on the ASV system should be explored.

Additionally, methods for optimized target selection can be explored, such as optimization techniques based on Bayesian hill climbing [99]. Another future direction is toward exploring and analyzing the weaknesses of ASV system in terms of the evaluation metrics. The behavior of various output parameters of the ASV system which affect the performance measures, such as EER and t-DCF, should be explored w.r.t. various attacks. To that effect, the question "How can different attacking approaches impact the performance measures of the voice biometric system?" remains unanswered and is an open research problem.

Acknowledgments

The authors would like to thank the organizers of INTERSPEECH 2020 special session on attacker's perspective, and the authorities of DA-IICT Gandhinagar, India, for their support. They would also like to thank Prof. (Dr) Rodrigo Capobianco Guido (São Paulo State University, Brazil) for his kind help and cooperation in going through the entire manuscript for English-language corrections. The second author would like to thank Dr Rohan Kumar Das (of NUS Singapore) for the interesting discussions on attacker's perspective during APSIPA ASC 2019.

References

[1]   Rosenberg A.E. 'Automatic SPEAKER verification: a review'. *Proceedings of the IEEE*. 1976;64(4):475–87.

[2]   Malik K.M., Malik H., Baumann R. 'Towards vulnerability analysis of voice-driven interfaces and countermeasures for replay attacks'. *IEEE Conference on Multimedia Information Processing and Retrieval (MIPR)*. San Jose: California, USA; 2019. pp. 523–8.

[3]   Wu Z., Evans N., Kinnunen T., Yamagishi J., Alegre F., Li H. 'Spoofing and countermeasures for SPEAKER verification: a survey'. *Speech Communication*. 2015;66:130–53.

[4]   Stylianou Y., Cappé O., Moulines E. 'Continuous probabilistic transform for voice conversion'. *IEEE Transactions on Speech and Audio Processing*. 1998;6(2):131–42.

[5]   Stylianou Y. 'Voice transformation: A survey'. *International Conference on Acoustics, Speech, and Signal Processing (ICASSP)*. Taipei, Taiwan; 2009. pp. 3585–8.

[6]   Zen H., Tokuda K., Black A.W. 'Statistical parametric speech synthesis'. *Speech Communication*. 2009;51(11):1039–64.

[7]   De Leon P.L., Pucher M., Yamagishi J., Hernaez I., Saratxaga I. 'Evaluation of SPEAKER verification security and detection of HMM-based synthetic speech'. *IEEE Transactions on Audio, Speech, and Language Processing*. 2012;20(8):2280–90.

[8]   Alegre F., Janicki A., Evans N. 'Re-assessing the threat of replay spoofing attacks against automatic speaker verification'. *International Conference of the Biometrics Special Interest Group (BIOSIG)*. Darmstadt, Germany; 2014. pp. 1–6.

[9]   Paul A., Das R.K., Sinha R., Mahadeva Prasanna S.R. 'Countermeasure to handle replay attacks in practical speaker verification systems'. *2016 International Conference on Signal Processing and Communications (SPCOM)*. Bengaluru, India: IISc; 12–15 Jun 2016. pp. 1–5.

[10] Prajapati G.P., Kamble M.R., Patil H.A. 'Energy separation based features for replay spoof detection for voice assistant'. *Accepted in European Signal Processing Conference (EUSIPCO)*. Netherlands: Amsterdam; 2021.

[11] Lau Y.W., Wagner M., Tran D. 'Vulnerability of speaker verification to voice mimicking'. International Symposium on *Intelligent Multimedia, Video, and Speech Processing*; Hong Kong; 2004. pp. 145–8.

[12] Evans N.W.D., Kinnunen T., Yamagishi J. 'Spoofing and countermeasures for automatic speaker verification'. INTERSPEECH, 25–29 August; 2013. pp. 925–9.

[13] Kersta L.G., Colangelo J.A. 'Spectrographic speech patterns of identical twins'. *The Journal of the Acoustical Society of America*. 1970;47(1A):58–9.

[14] Patil H.A., Parhi K.K. 'Variable length teager energy based mel cepstral features for identification of twins'. *International Conference on Pattern*

Recognition and Machine Intelligence; New Delhi, India, 16–20 December; 2009. pp. 525–30.

[15] Hsbc reports high trust levels in biometric tech as twins spoof its voice id system [online]. Biometric Technology Today, 2017(6):12, 2017. ISSN 0969-4765. https://doi.org/10.1016/S0969-4765(17)30119-4. Available from http://www.sciencedirect.com/science/article/pii/S0969476517301194 [Accessed 31 Jan 2020].

[16] Editorial Team. *Twins fool HSBC voice biometrics - BBC* [online]. 2017. Available from https://www.finextra.com/newsarticle/30594/twins-fool-hsbc-voice-biometrics-bbc [Accessed 31 Jan 2020].

[17] Hunt A.J., Black A.W. 'Unit selection in a concatenative speech synthesis system using a large speech database'. *1996 IEEE International Conference on Acoustics, Speech, and Signal Processing Conference Proceedings (ICASSP)*, volume 1; Georgia, USA, 7–10 May; 1996. pp. 373–6.

[18] Soong F.K., Qian Y., Yan Z.-J. 'A unified trajectory tiling approach to high quality speech rendering'. *IEEE Transactions on Audio, Speech, and Language Processing*. 2012;21(2):280–90.

[19] Saito Y., Takamichi S., Saruwatari H. 'Statistical parametric speech synthesis incorporating generative adversarial networks'. *IEEE/ACM Transactions on Audio, Speech, and Language Processing*. 2017;26(1):84–96.

[20] Kinnunen T., Wu Z.-Z., Lee K.A., Sedlak F., Chng E.S., Li H. 'Vulnerability of speaker verification systems against voice conversion spoofing attacks: The case of telephone speech'. *2012 IEEE International Conference on Acoustics, Speech and Signal Processing* (ICASSP); Kyoto, Japan, 25–30 March; 2012. pp. 4401–4.

[21] Bonastre J.-F., Matrouf D., Fredouille C. *Interspeech* [online]. 2007. Available from https://hal.archives-ouvertes.fr/hal-02157147.

[22] Abe M., Nakamura S., Shikano K., Kuwabara H. 'Voice conversion through vector quantization'. *Journal of the Acoustical Society of Japan*. 1990;11(2):71–6.

[23] Erro D., Moreno A. 'Weighted frequency warping for voice conversion'. *Eighth Annual Conference of the International Speech Communication Association*; Antwerp, Belgium, 27–31 August; 2007.

[24] Lindberg J., Blomberg M. 'Vulnerability in speaker verification-a study of technical impostor techniques'. *Sixth European Conference on Speech Communication and Technology*; Budapest, Hungary, 5–9 September; 1999.

[25] Villalba J., Lleida E. 'Speaker verification performance degradation against spoofing and tampering attacks'. FALA workshop; Vigo, Spain; 2010. pp. 131–4.

[26] Villalba J., Lleida E. 'Detecting replay attacks from far-field recordings on speaker verification systems'. European Workshop on Biometrics and Identity Management; Brandenburg, Germany; 2011. pp. 274–85.

[27] Zhizheng W., Kinnunen T., Evans N., *et al.* 'Asvspoof 2015: The first automatic speaker verification spoofing and countermeasures challenge'. *Sixteenth*

Annual Conference of the International Speech Communication Association; Dresden, Germany, 6–10 September; 2015. pp. 2037–41.

[28]   Novoselov S., Kozlov A., Lavrentyeva G., Simonchik K., Shchemelinin V. 'STC anti-spoofing systems for the ASVSpoof 2015 challenge'. *2016 IEEE International Conference on Acoustics, Speech and Signal Processing (ICASSP)*; Shanghai, China, 20–25 March; 2016. pp. 5475–9.

[29]   Wester M., Wu Z., Yamagishi J. 'Human vs machine spoofing detection on wideband and narrowband data'. *Sixteenth Annual Conference of the International Speech Communication Association*; Dresden, Germany, 6–10 September; 2015. pp. 2047–51.

[30]   Wang L., Yoshida Y., Kawakami Y., Nakagawa S. 'Relative phase information for detecting human speech and spoofed speech'. *Sixteenth Annual Conference of the International Speech Communication Association*; Dresden, Germany, 6–10 September; 2015. pp. 2092–6.

[31]   Liu Y., Tian Y., He L., Liu J., Johnson M.T. 'Simultaneous utilization of spectral magnitude and phase information to extract supervectors for speaker verification anti-spoofing'. *Sixteenth Annual Conference of the International Speech Communication Association*; Dresden, Germany, 6–10 September; 2015. pp. 2082–6.

[32]   Xiao X., Tian X., Du S., Xu H., Chng E.S., Li H. 'Spoofing speech detection using high dimensional magnitude and phase features: The ntu approach for ASVspoof 2015 challenge'. *Sixteenth Annual Conference of the International Speech Communication Association*; Dresden, Germany, 6–10 September; 2015. pp. 2052–6.

[33]   Wu Z., Yamagishi J., Kinnunen T., *et al.* 'Asvspoof: The automatic SPEAKER verification spoofing and countermeasures challenge'. *IEEE Journal of Selected Topics in Signal Processing*. 2017;11(4):588–604.

[34]   Font R., Espín J.M., Cano M.J. 'Experimental analysis of features for replay attack detection-results on the ASVSpoof 2017 challenge'. INTERSPEECH; Stockholm, Sweden, 20–24 August; 2017. pp. 7–11.

[35]   Witkowski M., Kacprzak S., Zelasko P., Kowalczyk K., Galka J. 'Audio replay attack detection using high-frequency features'. INTERSPEECH; Stockholm, Sweden, 20–24 August; 2017. pp. 27–31.

[36]   Wang X., Xiao Y., Zhu X. 'Feature selection based on CQCCs for automatic speaker verification spoofing'. INTERSPEECH; Stockholm, Sweden, 20–24 August; 2017. pp. 32–6.

[37]   Kamble M.R., Sailor H.B., Patil H.A., Li H. 'Advances in anti-spoofing: From the perspective of ASVspoof challenges'. *APSIPA Transactions on Signal and Information Processing*. 2020;9(Ed. 2).

[38]   *NIST Speaker Recognition Evaluation* [online]. 2019. Available from https://sre.nist.gov/ [Accessed 05 Feb 2020].

[39]   *Voice privacy challenge* [online]. 2020. Available from https://www.voicepri vacychallenge.org/ [Accessed 31 Jan 2020].

[40]   *Short-durationspeaker verification challenge* [online]. 2020. Available from https://sdsvc.github.io/ [Accessed 31 Jan 2020].

[41] *Far-field speaker verification challenge* [online]. Available from http://2020. ffsvc.org/ [Accessed 31 Jan 2020].

[42] Das R.K., Tian X., Kinnunen T., Li H. *The attacker's perspective on automatic speaker verification: An overview*. INTERSPEECH, 25–29 October. 2020. Available from https://arxiv.org/abs/2004.08849 [Accessed 14 May 2020].

[43] Kocher P., Jaffe J., Jun B., Rohatgi P. 'Introduction to differential power analysis'. *Journal of Cryptographic Engineering*. 2011;1(1):5–27.

[44] Kocher P., Jaffe J., Jun B. 'Differential power analysis'. Annual International Cryptology Conference; Santa Barbara, California, USA, 15–19 Aug; 1999. pp. 388–97.

[45] Akishita T., Takagi T. 'Zero-value point attacks on elliptic curve cryptosystem'. *International Conference on Information Security*; Bristol, United Kingdom; 2003. pp. 218–33.

[46] Cisse M., Adi Y., Neverova N., Keshet J. *Houdini: Fooling Deep Structured Prediction Models* [online]. arXiv preprint arXiv:1707.05373. 2017 [Accessed 31 Jan 2020].

[47] Szegedy C., Zaremba W., Sutskever I. *Intriguing Properties of Neural Networks* [online]. arXiv preprint arXiv:1312.6199. 2013 [Accessed 22 Jun 2020].

[48] Carlini N., Wagner D. 'Audio adversarial examples: Targeted attacks on speech-to-text'. IEEE Security and Privacy Workshops (*SPW*); San Francisco, 24 May; 2018. pp. 1–7.

[49] Papernot N., McDaniel P., Goodfellow I., Jha S., Berkay Celik Z., Swami A. 'Practical black-box attacks against machine learning'. Proceedings of the *2017 ACM on Asia Conference on Computer and Communications Security*, 2–6 Apr; 2017. pp. 506–19.

[50] Chen G., Sen Chen L.F., Du X., Zhao Z., Song F., Liu Y. *Who Is Real Bob? Adversarial Attacks on Speaker Recognition Systems* [online]. arXiv preprint arXiv:1911.01840. 2019 [Accessed 14 May 2020].

[51] Tian X., Das R.K., Li H. Black-Box Attacks on Automatic Speaker Verification Using Feedback-Controlled Voice Conversion [online]. arXiv preprint arXiv:1909.07655. 2019 [Accessed 05 Feb 2020].

[52] Papernot N., McDaniel P., Goodfellow I., Jha S., Berkay Celik Z., Swami A. 'Practical black-box attacks against machine learning'. Proceedings of the 2017 *ACM on Asia conference on computer and communications security*; 2017. pp. 506–19.

[53] Kreuk F., Adi Y., Cisse M., Keshet J. 'Fooling end-to-end speaker verification with adversarial examples'. *IEEE International Conference on Acoustics, Speech and Signal Processing* (ICASSP); Calgary, Canada, 15–20 April; 2018. pp. 1962–6.

[54] Goodfellow I., Pouget-Abadie J., Mirza M., *et al.* 'Generative adversarial nets'. *Advances in Neural Information Processing Systems* (NIPS); Montreal, Canada; 2014a. pp. 2672–80.

[55]   Goodfellow I.J., Shlens J., Szegedy C. *Explaining and Harnessing Adversarial Examples* [online]. arXiv preprint arXiv:1412.6572. 2014b [Accessed 22 Jun 2020].

[56]   Madry A., Makelov A., Schmidt L., Tsipras D., Vladu A. *Towards Deep Learning Models Resistant to Adversarial Attacks* [online]. arXiv preprint arXiv:1706.06083. 2017 [Accessed 13 Sep 2020].

[57]   Liu S., Wu H., Lee H.-yi., Meng H. 'Adversarial attacks on spoofing countermeasures of automatic speaker verification'. *IEEE Automatic Speech Recognition and Understanding Workshop* (ASRU); Singapore; 2019. pp. 312–9.

[58]   Document iso/iec 24745:2011, information technology—security techniques—biometric information protection. 'ISO/IEC JTCI SC27 security techniques'. 2011.

[59]   Gomez-Barrero M., Galbally J., Rathgeb C., Busch C. 'General framework to evaluate unlinkability in biometric template protection systems'. *IEEE Transactions on Information Forensics and Security*. 2017;13(6):1406–20.

[60]   Srivastava B.M.L., Bellet A., Tommasi M., Vincent E. *Privacy-preserving adversarial representation learning in asr: Reality or illusion* [online]? arXiv preprint arXiv. 2019;1911.04913. 2019 [Accessed 9 Aug 2020].

[61]   Nautsch A., Jiménez A., Treiber A., *et al.* 'Preserving privacy in SPEAKER and speech characterisation'. *Computer Speech & Language*. 2019;58(2):441–80.

[62]   Lorenzo-Trueba J., Fang F., Wang X., Echizen I., Yamagishi J., Kinnunen T. 'Can we steal your vocal identity from the internet?: Initial investigation of cloning obama's voice using gan, wavenet and low-quality found data'. *arXiv preprint arXiv:1803.00860*. 2018.

[63]   Vestman V., Soomro B., Kanervisto A., Hautamäki V., Kinnunen T. 'Who do i sound like? Showcasing speaker recognition technology by youtube voice search'. IEEE International Conference on Acoustics, Speech and Signal Processing (ICASSP); Brighton, United Kingdom, 12–17 May; 2019. pp. 5781–5.

[64]   Vestman V., Kinnunen T., González Hautamäki R., Sahidullah M. 'Voice mimicry attacks assisted by automatic SPEAKER verification'. *Computer Speech & Language*. 2020;59(1):36–54.

[65]   Gupta P., Patil H.A., Guido R.C. 'A Survey on Attackers' Perspectives in Automatic Speaker Verification (ASV) Systems'. Submitted to International Conference on Acoustics, Speech and Signal Processing (ICASSP); Ontario, Canada, 6–11 June; 2021.

[66]   Li H., Patil H.A., Kamble M.R. 'Tutorial on spoofing attack of speaker recognition'. Asia-Pacific Signal and Information Processing Association, Annual Summit and Conf. (APSIPA-ASC); Kuala Lumpur, Malaysia, 12–15 Dec; 2017.

[67]   Doddington G., Liggett W., Martin A., Przybocki M., Reynolds D. 'Sheep, goats, lambs and wolves: a statistical analysis of SPEAKER performance in the NIST 1998 SPEAKER recognition evaluation. technical report, National insT of standards and technology (NIST) Gaithersburg MD'. 1998.

[68] Kumar R., Jovanovic P., Burleson W., Polian I. 'Parametric trojans for fault-injection attacks on cryptographic hardware'. In IEEE, Workshop on Fault Diagnosis and Tolerance in Cryptography; Busan, South Korea, 23 Sep; 2014. pp. 18–28.

[69] Barenghi A., Breveglieri L., Koren I., Naccache D. 'Fault injection attacks on cryptographic devices: theory, practice, and countermeasures'. *Proceedings of the IEEE*. 2012;100(11):3056–76.

[70] Li Z., Shi C., Xie Y., Liu J., Yuan B., Chen Y. 'Practical adversarial attacks against speaker recognition systems'. *Proceedings of the 21st International Workshop on Mobile Computing Systems and Applications*; Austin TX USA March; 2020. pp. 9–14.

[71] Xie Y., Shi C., Li Z., Liu J., Chen Y., Yuan B. 'Real-time, universal, and robust adversarial attacks against speaker recognition systems'. IEEE International Conference on Acoustics, Speech and Signal Processing (ICASSP); Barcelona, Spain, 4–8 May; 2020. pp. 1738–42.

[72] Marras M., Korus P., Memon N.D., Fenu G. 'Adversarial optimization for dictionary attacks on speaker verification'. INTERSPEECH; Graz, Austria, 15–19 Sept; 2019. pp. 2913–7.

[73] Reynolds D., Andrews W., Campbell J., *et al.* The supersid project: Exploiting high-level information for high-accuracy speaker recognition. *IEEE International Conference on Acoustics, Speech, and Signal Processing Proceedings (ICASSP)*, volume 4; Hong Kong, China, 6–10 April; 2003. pp. IV–784.

[74] Patil H.A. *Speaker recognition in Indian languages: A feature based approach [PhD Thesis]*. Department of Electrical Engineering, Indian Institute of Technology (IIT) Kharagpur; 2005.

[75] Campbell J.P., Reynolds D.A., Dunn R.B. 'Fusing high-and low-level features for speaker recognition'. Eighth European Conference on Speech Communication and Technology; Geneva, Switzerland, 1–4 Sep; 2003.

[76] Campbell W.M., Campbell J.R., Reynolds D.A., Jones D.A., Leek T.R. 'High-level speaker verification with support vector machines'. IEEE International Conference on Acoustics, Speech, and Signal Processing (ICASSP), volume 1; Montreal, Quebec, Canada, 17–21 May; 2004. pp. I–73.

[77] Reynolds D.A. 'Automatic SPEAKER recognition: Current approaches and future trends'. *Speaker Verification: From Research to Reality*. 2001;5:14–15.

[78] Reynolds D., Campbell J.P., Campbell B., *et al.* Beyond cepstra: Exploiting high-level information in speaker recognition. Workshop on Multimodal User Authentication; Santa Barbara, California, 11–12 Dec; 2003. pp. 223–9.

[79] Alessandra Aparecida Paulino. *Contributions to biometric recognition: Matching identical twins and latent fingerprints [PhD Thesis]*. Department of Computer Science, Michigan State University; 2013.

[80] Patil H.A., Dutta P.K., Basu T.K. 'Effectiveness of LP-based features for identification of professional mimics in indian languages'. Int. Workshop on Multimodal User Authentication; MMUA, Toulouse, France; 2006.

[81] Gong Y., Poellabauer C. *An Overview of Vulnerabilities of Voice Controlled Systems* [online]. arXiv preprint arXiv:1803.09156. 2018 [Accessed 21 April 2020].

[82] Wu H., Liu S., Meng H., Lee H.-yi. 'Defense against adversarial attacks on spoofing countermeasures of asv'. *IEEE International Conference on Acoustics, Speech and Signal Processing* (ICASSP); Barcelona, Spain, 4–8 May; 2020. pp. 6564–8.

[83] Miyato T., Maeda S.-ichi., Koyama M., Nakae K., Ishii S. *Distributional smoothing with virtual adversarial training* [online]. arXiv preprint arXiv:1507.00677. 2015 [Accessed 14 May 2020].

[84] Xu W., Evans D., Qi Y. *Feature squeezing: Detecting adversarial examples in deep neural networks. Proceedings network and distributed system security symposium* [online]. arXiv preprint arXiv:1704.01155. 2017 [Accessed 14 May 2020].

[85] Wang Q., Guo P., Sun S., Xie L., Hansen J.H.L. 'Adversarial regularization for end-to-end robust speaker verification'. INTERSPEECH; Graz, Austria, Sep. 15–19; 2019. pp. 4010–4.

[86] Campbell J.P. 'Speaker recognition: a tutorial'. *Proceedings of the IEEE.* 1997;85(9):1437–62.

[87] Reynolds Douglas A. 'An overview of automatic speaker recognition technology'. IEEE International Conference on Acoustics, Speech, and Signal Processing (ICASSP), volume 4; Orlando, Florida, USA, May 13–17; 2002. pp. IV–4072.

[88] Cohen-Hadria A., Cartwright M., McFee B., Bello J.P. 'Voice anonymization in urban sound recordings'. *IEEE 29th International Workshop on Machine Learning for Signal Processing* (MLSP); Pittsburgh, PA, USA, 13–16 Oct; 2019. pp. 1–6.

[89] Gontier F., Lagrange M., Lavandier C., Petiot J.-F. 'Privacy aware acoustic scene synthesis using deep spectral feature inversion'. IEEE International Conference on Acoustics, Speech and Signal Processing (ICASSP); Virtual Barcelona, Spain, 4–8 May; 2020. pp. 886–90.

[90] Pathak M.A., Raj B., Rane S.D., Smaragdis P. 'Privacy-preserving speech processing: Cryptographic and string-matching frameworks show promise'. *IEEE Signal Processing Magazine.* 2013;30(2):62–74.

[91] Smaragdis P., Shashanka M. 'A framework for secure speech recognition'. *IEEE Transactions on Audio, Speech and Language Processing.* 2007;15(4):1404–13.

[92] Zhang S.-X., Gong Y., Yu D. 'Encrypted speech recognition using deep polynomial networks'. *IEEE International Conference on Acoustics, Speech and Signal Processing* (ICASSP); Brighton, UK, 12–17 May; 2019. pp. 5691–5.

[93] Brasser F., Frassetto T., Riedhammer K., Sadeghi A.-R., Schneider T., Weinert C. 'Voiceguard: Secure and private speech processing'. INTERSPEECH; Hyderabad, India, 2–6 Sep; 2018. pp. 1303–7.

[94] Leroy D., Coucke A., Lavril T., Gisselbrecht T., Dureau J. 'Federated learning for keyword spotting'. *IEEE International Conference on Acoustics, Speech*

and Signal Processing (ICASSP); Brighton, UK, 12–17 May; 2019. pp. 6341–5.

[95] Geiping J., Bauermeister H., Dröge H., Moeller M. *Inverting gradients–how easy is it to break privacy in federated learning?* arXiv preprint arXiv:2003.14053. 2020 [Accessed 28 Jun 2020].

[96] Shah N.J., Parmar M., Shah N., Patil H.A. 'Novel mmse discogan for cross-domain whisper-to-speech conversion'. *Machine Learning in Speech and Language Processing* (MLSLP) Workshop; Hyderabad, India, 17–20 Sep; 2018. pp. 1–3.

[97] Soni M.H., Shah N., Patil H.A. 'Time-frequency masking-based speech enhancement using generative adversarial network'. *2018 IEEE International Conference on Acoustics, Speech and Signal Processing* (ICASSP); Calgary, Alberta, Canada, 15–20 Apr; 2018. pp. 5039–43.

[98] Patel M., Parmar M., Doshi S., Shah N., Patil H.A. 'Novel inception-GAN for whisper-to-normal speech conversion'. The ISCA Speech Synthesis Workshop (SSW); Viennna, Austria, 20–22 Sep; 2019.

[99] Gámez J.A., Mateo J.L., Puerta J.M. 'Learning Bayesian networks by Hill climbing: Efficient methods based on progressive restriction of the neighborhood'. *Data Mining and Knowledge Discovery*. 2011;22(1–2):106–48.

Chapter 4

Voice biometrics: privacy in paralinguistic and extralinguistic tasks for health applications

Francisco Teixeira[1], Alberto Abad[1], Isabel Trancoso[1], and Bhiksha Raj[2]

The widespread use of cloud computing applications has created a society-wide debate on how user privacy is handled by online service providers. Regulations such as the European Union's General Data Protection Regulation have put forward restrictions on how such services are allowed to handle user data. The field of privacy-preserving machine learning (PPML) is a response to this issue that aims to develop secure classifiers for remote prediction, where both the client's data and the server's model are kept private. This is particularly relevant in the case of speech and concerns not only the linguistic contents but also the paralinguistic and extralinguistic information that may be extracted from the speech signal. In this chapter, we provide a brief overview of the current state of the art in paralinguistic and extralinguistic tasks for a major application area in terms of privacy concerns – health, along with an introduction to cryptographic methods commonly used in PPML. These will lay the groundwork for the review of the state of the art of privacy in paralinguistic and extralinguistic tasks for health applications. With this chapter we hope to raise awareness to the problem of preserving privacy in this type of tasks and provide an initial background for those who aim to contribute to this topic.

4.1 Introduction

The widespread use of mobile devices, cloud-based applications, social media platforms and, more recently, voice-based virtual assistants, has given companies and researchers unprecedented access to data. In turn, this has allowed the development of highly accurate predictive machine learning (ML) models. Cloud-based platforms leverage such models to provide users with (remote) Machine Learning as a

[1]INESC-ID, Instituto Superior Técnico, University of Lisbon, Portugal
[2]Language Technologies Institute, Carnegie Mellon University, Pittsburgh, Pennsylvania

Service (MLaaS) applications that automate time-consuming tasks (e.g. document translation, transcribing speech, image labeling) and help users perform everyday tasks (e.g. voice-based virtual assistants).

Most MLaaS applications require an input from the user to be sent to a remote server for processing, be it a query, text, image, video and speech or other biometric signals. However, this setting creates two potential privacy issues: first, the afore-mentioned inputs can be used to determine information about the user, including their preferences, personality traits, mood, health, political opinions, among many others; second, depending on the task at hand, the output of the ML model can itself be of a sensitive nature – for example the user sends a biometric signal to a service provider to determine whether they present symptoms of a disease. Yet, in most cases, users are given few to no privacy guarantees for either their data or the information that may be extracted by processing it. Even though some concerns may be attenuated through service-user agreements, conformance to these agreements heavily depends on the service provider's, sometimes questionable, accountabil-ity. Moreover, the alternative of performing the computation directly on the user's device is unattractive to many service providers, since it would require abdicating the privacy of their own model, defeating the purpose of the service. The recent European Union's General Data Protection Regulation [1] is a strong indicator of the growing societal awareness to the problem of data misuse, and of the pressing need to develop secure predictive methods that take user privacy into account.

Among other biometric signals, speech stands out for the amount of informa-tion it carries about a speaker. Speech includes not only linguistic information but also paralinguistic and extralinguistic information such as age, gender, ethnicity, personality traits, emotional state, and even information regarding the physical and mental health of a speaker. All this information can be considered private and thus it should be protected at all costs. In fact, we live in a world where speech data and the information that may be extracted from it may be legally regarded as personable identifiable information [1–3].

For these reasons, and with the help of breakthroughs in homomorphic encryp-tion (HE) and secure multiparty computation (MPC), interest in PPML frameworks (cf. Figure 4.1) has grown exponentially in recent years. Works such as Cryptonets [4], MiniONN [5], Chameleon [6], Gazelle [7], ABY [8], and Delphi [9] have pushed this field forward to a point where private inference over ML models for benchmark datasets such as MNIST [10] and CIFAR-10 [11] does not suffer any relevant loss of accuracy when compared to the original *in-the-clear* models. Moreover, open-source libraries such as HELib [12], SEAL [13], ABY [14], and MP-SPDZ [15, 16] have helped make reproducible research and have allowed non-cryptographers to contribute to this topic with their own expert knowledge in areas such as speech processing and ML.

However, in spite of the above, privacy-preserving speech processing is still in its early days. The speech community has only recently started to become aware of the problem of privacy in speech [2, 3]. Nonetheless, this is a rapidly growing field, as shown by works for privacy-preserving speaker verification [17, 18], query-by-example speech search [19], and for paralinguistic [20] and extralinguistic [21,

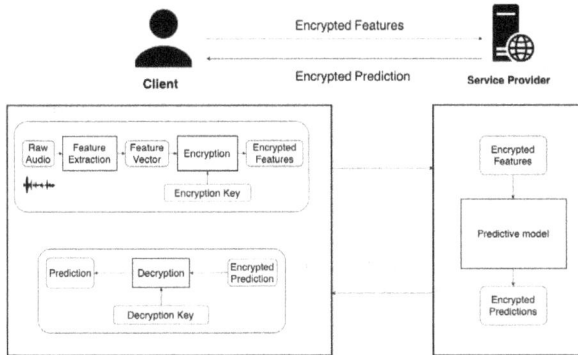

Figure 4.1   Example of a privacy-preserving machine learning framework

22] tasks. In this chapter, we will focus mainly on the latter and, in particular, on health-related tasks, as these are among the most relevant in terms of privacy concerns.

This chapter is structured as follows: Section 4.2 briefly reviews the state of the art in paralinguistic and extralinguistic health related tasks; Section 4.3 includes an overview of the main cryptography concepts relevant for privately performing such tasks; Section 4.4 describes the state of the art in privacy-preserving paralinguistic and extralinguistic tasks and illustrates the potential and the challenges presented by such tasks. Finally, Section 4.5 includes some conclusions about the topics discussed in the chapter.

4.2   Paralinguistic and extralinguistic tasks

Although the speech research community has long been actively studying speaker recognition (and later language and accent recognition), the interest in other non-linguistic tasks is much more recent and has been fueled by a series of challenges organized mainly at Interspeech conferences, since 2009. Collectively known as ComParE (Computational Paralinguistics ChallengE)[1], these challenges have covered the automatic detection of a wide range of totally different traits, from intoxication to emotions, cognitive load, social signals, personality traits, and degree of non-nativeness, just to name a few. Another series of challenges on paralinguistic tasks is the Audio/Visual Emotion Challenge and Workshop (AVEC), held yearly at the Association for Computing Machinery's (ACM)'s Multimedia conferences since 2011 [23]. This series of challenges is mainly focused on multimodal emotion

[1]www.compare.openaudio.eu

recognition and, as the ComParE challenges, has also helped to raise interest in paralinguistic tasks.

A major application area for this type of tasks is health. Indeed, many health disorders affect speech at different stages of production including planning, respiration, phonation, and articulation. Diseases affecting organs shared by speech production will therefore affect speech as well. Consequently, a large number of disorders can be detected through speech.

The potential of speech to act as a biomarker for speech disorders has bolstered several studies on their automatic detection based on powerful ML classifiers that either operate on the basis of features extracted from the speech signal or directly on the signal itself (e.g. end-to-end deep learning networks). The fact that speech is ubiquitous and can be acquired non-intrusively makes it an inexpensive modality that may be used to identify high likelihoods of the presence of diseases, and the results of such screening tests may act as alerts for users to seek medical assistance. Moreover, speech may be used by clinicians and patients in many scenarios: in clinical facilities, or even at the patients' homes. It may also allow to remotely monitor the progress of a disease in order to adapt medication and support, which may be specially advantageous, given the usually long interval between specialized medical exams.

4.2.1   Speech-affecting diseases

For the sake of illustrating the impact of diseases in the non-linguistic aspects of speech production, we have selected two speech-affecting diseases: obstructive sleep apnea (OSA) and Parkinson's disease (PD). The first one was selected as a representative of breathing disorders, and the second one as a representative of neurodegenerative disorders. A third type of speech-affecting diseases, mood disorders (such as depression, for instance), will not be mentioned here, as typically paralinguistic cues may be explored in parallel with linguistic cues. In fact, whereas depressed speech is often described as dull, monotone, monoloud, lifeless and metallic, the analysis of its linguistic content, which often includes descriptions of traumatic events or topics, such as alcoholism and drugs, tends to give better results than paralinguistic/extralinguistic cues do by themselves [24]. The same can be said about Alzheimer's disease and other types of dementia, where linguistic and paralinguistic cues may both be very strong indicators of the presence of this type of diseases [25, 26]. Although the analysis of the text that is automatically produced by a speech recognition system may reveal such cues, these linguistically oriented works fall out of the scope of this chapter

- *Obstructive sleep apnea*: Patients suffering from OSA commonly show articulatory anomalies (caused by hypotonus or lack of regulated innervations to the breathing musculature, leading to slurred speech), phonation anomalies (consequence of a larynx inflammation caused by snoring that could affect the vocal folds), and abnormal coupling of the vocal tract with the nasal cavity, which is present even in non-nasal sounds [27].

- *Parkinson's disease*: Speech impairments are one of the earliest manifestations of PD, being present in 70%–90% of patients with PD. The most common of these disturbances are excess of tremor, reduced loudness, monotonicity, hoarseness, and imprecise articulation. Speech affected by this set of disturbances is called dysarthric speech. This type of impaired speech is also prevalent in other neurodegenerative diseases such as amyotrophic lateral sclerosis (or Lou Gehrig's disease) and Huntington's disease [28–31].

4.2.2 Methods

Some of the symptoms described above are visible in prosodic, voice quality, and spectral features that may be automatically extracted from the acoustic signal (e.g. pitch, energy, resonance frequencies, jitter, harmonicity, speech rate, pause duration, etc.) [32, 33]. Hence, the dominant approach to the detection of most speech-affecting diseases is based on sets of such knowledge-based (KB) features (e.g. Botelho *et al.* [27] for OSA and Pompili *et al.* [34] for PD). One justification for this is the fact that KB features are more easily explainable, thus making it possible to create bridges between data scientists and healthcare professionals. On the other hand, temporal and financial constraints, ethical issues and patient privacy laws make the acquisition of data a difficult and lengthy task, which often results in very small datasets. Using KB features helps compensate this lack of data by allowing the use of simpler classifiers that require smaller amounts of training data.

Several works have also been successful in the detection of speech-affecting diseases through the use of convolutional neural networks (CNNs), sometimes in combination with recurrent layers such as long short-term memory layers, and low-level descriptors (LLDs) as inputs (e.g. Mel frequency cepstral coefficients (MFCCs), Mel Filterbank Energies, spectrograms, and even raw audio). This type of model, sometimes called end-to-end, is able to create internal representations that encapsulate the most relevant information for the task at hand, as opposed to KB features that may not represent the more subtle symptoms of the disease and may even include unnecessary or redundant information. Speaker representations like *Gaussian Supervectors*, *i-vectors* and *x-vectors* have also been considered for disease classification. The intuition that speaker representations should contain information about existing speech disorders has been proven true by several works, including Arias-Vergara *et al.*, Hauptman *et al.*, Laaridh *et al.*, and Moro-Velazquez *et al.* [35–38] for the detection and assessment of PD, Perero-Codosero *et al.* [39] for the detection of OSA, and Zargarbashi *et al.* [40] for the detection of Alzheimer's disease.

Finally, some works have also applied Bags-of-Audio-Words (BoAWs) to perform disease classification. Similar to Bags-of-Words (BOW) for text classification, where a BOW vector represents the distribution of words according to a predefined vocabulary, BoAW vectors represent audio segments as distributions of LLDs from a codebook. This codebook is generated using LLDs extracted from the training data. This type of approach has been used to determine the source location of

OSA-related snoring and for the assessment of mood disorders, such as depression and post-traumatic stress disorder [41, 42].

4.3   Cryptographic primitives and MPC for PPML

Mining speech signals for health-related cues raises very important privacy concerns. The possibility of processing speech while protecting an individual's privacy will allow, for instance, a recording of someone's voice to be processed in a remote server for medical analysis. Combining speech processing with cryptographic techniques has been on the forefront of this ambitious goal. For this reason, in this section we will give a brief introduction to common cryptographic methods used for privacy-preserving inference, and present a short overview of the state of the art in PPML.

It is important to note that we will address neither differential privacy nor federated learning techniques in this chapter. Although extremely important for decentralized learning of ML models, these techniques are complementary to the privacy-preserving inference techniques described here, but cannot function as alternatives. Another line of research we will not cover is PPML based on secure enclaves, particularly Intel's Security Guard Extensions (SGX) enclave. Intel's SGX has been shown to be liable to several attacks, and, so far, its security remains unreliable [43–45]. For in-depth descriptions of these techniques we direct the reader to Dwork, Konečný *et al.*, and Costan *et al.* [46–48].

4.3.1   Homomorphic encryption

HE is a type of cryptosystem in which certain operations performed on *ciphertexts* (i.e. encrypted values) are *homomorphic* with regard to the *plaintexts* (i.e. unencrypted values). In other words, considering the encryption of a value x, $E(x)$ and of a value y, $E(y)$, if a homomorphic operation is performed on the two ciphertexts, the result of this operation will correspond to the equivalent unencrypted operation of the two values, as follows:

$$E(x) \otimes E(y) = E(x \times y)$$
$$E(x) \oplus E(y) = E(x + y)$$

(4.1)

HE can be roughly divided into three categories: partially homomorphic encryption (PHE), somewhat homomorphic encryption (SHE), and fully homomorphic encryption (FHE). PHE schemes are limited by the type of operation allowed, while SHE schemes are constrained in the number of times each operation can be performed over a ciphertext. In FHE schemes, neither of these restrictions apply, and all the allowed operations (usually additions and multiplications) can be performed an unlimited number of times.

PHE schemes such as Paillier [49], and additively homomorphic and multiplicatively homomorphic cryptosystems such as RSA [50] and ElGamal [51], have been extensively used in the literature for privacy-preserving applications.

The first FHE scheme, proposed by Craig Gentry in 2009 [52], allowed for unlimited operations to be performed over ciphertexts, but was highly inefficient. Since then, FHE schemes have been constantly improved and there are currently much faster implementations [53].

However, these implementations are still computationally heavy and, for most applications, it is not necessary to perform an unlimited number of operations over ciphertexts. In fact, for most applications, the user knows beforehand the number of operations that will be performed. In these cases, SHE schemes such as Brakerski-Gentry-Vaikuntanathan (BGV) [54], Brakerski-Fan-Vercauteren (BFV) [55], and more recently, Cheon-Kim-Kim-Song (CKKS) [56], allow the user to select the scheme's encryption parameters in such a way that it is only possible to perform a limited amount of operations as a trade-off with efficiency. In SHE schemes, to perform more operations it is necessary to use larger encryption parameters, which results in more expensive computations. To compensate for this limitation, SHE cryptosystems such as BGV, BFV, and CKKS encompass *batching* techniques that allow several messages to be encrypted in the same ciphertext and thus to be operated as single instruction multiple data (SIMD), effectively reducing the scheme's computational cost [54]. These schemes and their variations have recently started to be the prominent choice for PPML applications [4, 7, 57, 58].

4.3.2 Oblivious transfer

Oblivious transfer (OT) is a cryptographic primitive that allows two parties, a sender S and a receiver R, to exchange data in a private setting. In 1-out-of-n OT, S is in possession of n messages, one of which R is interested in, but does not want S to find out which, whereas S is willing to give R one of the messages, but does not want R to learn anything about the other messages [59, 60]. A particular case of this formulation, 1-out-of-2 OT, where there are only two messages for R to choose from, was first proposed by Rabin in 1981 [61]. OT optimizations such as *OT pre-computations* [62] and *OT extensions* [63] have to be highlighted due to the important role they play in improving the efficiency of OT-based MPC protocols.

4.3.3 Secure multiparty computation

MPC is an umbrella term for protocols designed to allow several parties to jointly, and interactively, compute a function over their data, while keeping all inputs private. Among others, MPC protocols include general protocols such as arithmetic and Boolean sharing (Ben-Or-Goldwasser-Wigderson (BGW) and Goldreich-Micali-Wigderson (GMW) protocols, respectively [64]) and Yao's garbled circuits (GCs) [65, 66], as well as protocols that combine HE and OT to perform specific functionalities [67]. In this section, we will describe MPC protocols for both Boolean and arithmetic functions followed by some necessary security definitions required for MPC applications.

Table 4.1 XOR gate truth tables

a	b	$z = a \oplus b$	a	b	$z = a \oplus b$
0	0	0	k_a^0	k_b^0	$\varepsilon_{k_a^0}(\varepsilon_{k_b^0}(k_z^0))$
0	1	1	k_a^0	k_b^1	$\varepsilon_{k_a^0}(\varepsilon_{k_b^1}(k_z^1))$
1	0	1	k_a^1	k_b^0	$\varepsilon_{k_a^1}(\varepsilon_{k_b^0}(k_z^1))$
1	1	0	k_a^1	k_b^1	$\varepsilon_{k_a^1}(\varepsilon_{k_b^1}(k_z^0))$
(a) Original truth table			**(b) Garbled truth table**		

4.3.3.1 Yao's GCs protocol

First proposed by Yao *et al.* in 1986 [66], GCs are a cryptographic construction that allows two parties, Alice and Bob, to jointly compute a function represented as a Boolean circuit, such that their inputs as well as any intermediate results are kept private and only the functions' output is revealed to one, or both parties. This construction requires each party to take a role, one must be the *garbler*, which we will assume is Alice, and a second party, in this case Bob, will be the *evaluator*.

Consider an XOR gate (cf. Table 4.1 a), which contains three wires, two inputs, and one output. For each of these wires, Alice chooses two random values, one for each bit, obtaining six encryption keys: $k_a^0, k_a^1, k_b^0, k_b^1, k_z^0, k_z^1$. Alice then uses the keys to encrypt the output of each row, using a symmetric encryption scheme, obtaining a garbled truth table (cf. Table 4.1 b). Now we require a way for Bob to *evaluate* the circuit without learning Alice's inputs. Since each bit is a random value, Bob will not be able to learn anything from them. However, the rows of the table need to be permuted so that Bob cannot correspond them with the rows of the original function. Consequently, Alice must also permute the rows of her GC table. Afterwards, Alice can send Bob the encrypted output column of the GC table, as well as her input, that is a random value and, as such, Bob will not be able to determine the bit it corresponds to. The next step in this process is for Bob to receive his own inputs from Alice. To this end, Alice and Bob perform an 1-out-of-2 OT. Therefore, Alice does not learn which input bit Bob selected, and Bob will only be able to decrypt the key corresponding to his chosen bit and will not be able to learn anything about the other key. Bob is now able to decrypt the output of the circuit, using his and Alice's input keys. Depending on the implementation, this process can occur in different ways. For simplicity we will assume that, after decryption, it is possible to distinguish the correctly decrypted value from the values decrypted using incorrect keys. Finally, Bob can either keep his output result or share it with Alice. For a more general circuit, the protocol defined above simply needs to be generalized for all gates in the circuit, taking into account that intermediate results will also be keys that will serve as inputs for the subsequent gates.

We can thus summarize Yao's GC protocol as follows [3]:

1. The *garbler* transforms the function *f* to be computed into a Boolean circuit and generates keys for all wires and gates of the circuit.

2. The *garbler* sends the permuted garbled tables and the keys corresponding to his/her inputs to the *evaluator*.
3. The *evaluator* obtains his/her inputs through OTs with the *garbler*, and evaluates each gate using his/her own keys, as well as the *garbler's* keys.
4. Finally, the *evaluator* reveals the output of the circuit and decides to share it or not with the *garbler* according to what was agreed by both parties beforehand.

Since it was first proposed, this protocol has been subject to several optimizations, including the *point-and-permute* optimization, which only requires one decryption per gate; the *half-gates* optimization which reduces the bandwidth required to compute AND gates [68, 69] and the *free-XOR* technique, which allows the computation of XOR gates without communication and at a very low computational cost [70]. Besides being able to model any function, one of the advantages of Yao's GC protocol is the fact that it requires a constant number of rounds of communication.

4.3.3.2 Secret sharing

Secret sharing is a family of protocols that allow two or more parties to interactively compute Boolean or arithmetic functions. In the two-party case, a secret share of a value x can be defined as

$$\langle x \rangle = \langle x \rangle_0 \oplus \langle x \rangle_1 \tag{4.2}$$

where $\langle x \rangle_0$ and $\langle x \rangle_1$ represent the random-looking shares held by party 0 and 1, respectively, and $\oplus$ represents either addition or XOR, depending on whether arithmetic (addition) or Boolean (XOR) sharing is being considered.

Boolean sharing

As in Yao's GC protocol, in Boolean sharing (also known as the GMW protocol, proposed by Goldreich *et al.* in 1987 [71]), the function to be computed is represented as a Boolean circuit of shares in $\mathbb{Z}_2$. In this protocol, due to their associative property, XOR gates can be computed locally. On the other hand, AND gates require more intricate constructions, such as Beaver's *Multiplication Triples* (*MTs*) [62]. Generated during the protocol's setup phase using OTs or HE, MTs are sets of three shared variables $\langle a \rangle^B$, $\langle b \rangle^B$, and $\langle c \rangle^B$ (where B represents a Boolean share) for which the relation $\langle c \rangle^B = \langle a \rangle^B \wedge \langle b \rangle^B$ holds. If one wants to compute an AND gate between two shared values $\langle x \rangle^B$ and $\langle y \rangle^B$, each party takes its shares of the MTs and computes $\langle e \rangle_i^B = \langle a \rangle_i^B \oplus \langle x \rangle_i^B$ and $\langle f \rangle_i^B = \langle b \rangle_i^B \oplus \langle y \rangle_i^B$. Each party then shares their result with the others, allowing every party to reconstruct e and f. Finally, each party i can then compute

$$\langle z \rangle_i^B = i \cdot e \cdot f \oplus f \cdot \langle a \rangle_i^B \oplus e \cdot \langle b \rangle_i^B \oplus \langle c \rangle_i^B \tag{4.3}$$

From the above it is easy to show that by adding the shares of each party, the final result will be $\langle z \rangle^B = \langle x \rangle^B \cdot \langle y \rangle^B$. GMW requires a variable number of communication rounds and has been shown to be efficient over low-latency networks [14, 60].

Arithmetic secret sharing

Arithmetic secret sharing (or the BGW protocol, proposed by Ben-Or *et al.* in 1988 [72]) can be seen as a generalization of the GMW protocol for integers in $\mathbb{Z}_{2^\ell}$. Just as in the GMW protocol, due to the associative property of addition, this operation can also be computed locally, without communication with the other party. In the case of multiplications, this protocol requires MTs of the form $\langle a \rangle^A$, $\langle b \rangle^A$, and $\langle c \rangle^A$, where $\langle c \rangle^A = \langle a \rangle^A \times \langle b \rangle^A$. In this protocol, each party sets its shares to $\langle e \rangle_i^A = \langle x \rangle_i^A - \langle a \rangle_i^A$ and $\langle f \rangle_i^A = \langle y \rangle_i^A - \langle b \rangle_i^A$ and exchanges the results with the other parties, so that each party holds e and f. The resulting share is given by [14]

$$\langle z \rangle_i^A = i \cdot e \cdot f + f \cdot \langle a \rangle_i^A + e \cdot \langle b \rangle_i^A + \langle c \rangle_i^A \tag{4.4}$$

Secret sharing schemes have the advantage of being much lighter computationally than HE. However, they require the online presence of all the parties involved in the computation, and usually have higher costs in terms of communication due to the multiple rounds of interaction.

4.3.3.3 Security models

In MPC it is necessary to define how an adversary will behave, to guarantee the security of the protocol. The most common security (or threat) models include the *honest-but-curious* (also called *semi-honest*) adversary model and the *malicious* adversary model.

The *honest-but-curious* model is the simplest model possible and it is considered to be sufficient for most applications [5–7, 14]. In this model, the adversary is assumed to follow the established protocol, but is also assumed to pry into data that is visible to him/her. In this way, there is no need to create additional safeguards outside of the protocol's inherent security, allowing for very efficient implementations. The *honest-but-curious* model is used in applications where both parties are trustworthy (e.g. interaction between hospitals or clinics and companies).

The *malicious* model drops the assumption that the adversary follows the protocol, thus requiring a series of additional proofs to ensure that each party is following the protocol. This is usually done through zero knowledge (ZK) proofs and cut-and-choose methods [73]. This threat model should be used in settings where parties do not trust each other (e.g. two competing companies that need to perform a computation over their private data). Although much more secure, this model significantly increases the computational cost of MPC protocols.

More complex models exist that take other assumptions into account – for instance, whether the adversary changes behavior during the protocol's execution; what is the maximum number of corrupted parties allowed before the protocol can no longer be executed securely. However, these fall out of the scope of this chapter and we instead redirect readers who would like to learn more about this topic to the works of Lindell [74, 75].

4.3.4 Distance-preserving hashing techniques

Locality-sensitive hashing (LSH) is a family of locality-sensitive hash functions $H(.)$ that project vectors into lower-dimensional spaces. If two vectors are close enough in the input space, they will hash to the same value, or "bucket," with high probability. On the other hand, if two vectors are far apart in the input space, they will be projected onto different buckets. More formally, given two vectors, x and y, an LSH function will be a map h from space $\mathbb{M}$ to space $\mathbb{S}$ such that, given a distance function d in the input space

$$h(x) = h(y) \text{ with high probability, if } d(x - y) \leq r$$
$$h(x) \neq h(y) \text{ with high probability, if } d(x - y) \geq cr \tag{4.5}$$

where r is the radius of interest and c a constant greater than 1. If two inputs have a distance smaller than r, there will be a a high probability that they will have the same hash, whereas if their distance is greater than cr there will be a high probability that they will fall into different buckets [76, 77]. Although LSH can be used to mask data, it does not provide any security guarantee. In the remainder of this section, we will detail two distance-preserving hash functions, secure binary embeddings (SBEs) and secure modular hashing (SMH), that provide information-theoretic security.

Secure binary embeddings

SBEs [78] are a type of LSH that uses band-quantized random projections to convert real-valued vectors into bit sequences, providing information-theoretic security guarantees. SBE is based on universal quantization, which redesigns the scalar quantization function to have non-contiguous quantization regions. For a vector x with L features, the SBE transformation is a random projection from $\mathbb{R}^N$ into $\mathbb{Z}_2$, as follows:

$$Q_{SBE}(x) = \lfloor \Delta^{-1}(Ax + w) \rfloor (\bmod\ 2) \tag{4.6}$$

where Δ^{-1} is a diagonal matrix that defines the precision parameters, $A \in \mathbb{R}^{L \times M} \sim N(0, 1)$ is a random matrix, and $w \in \mathbb{R}^M \sim \text{unif}[0, \Delta]$ is the additive dither; the size of the output hash vector is usually defined in terms of the number of input features as $M = L \times \text{mpc}$, where *mpc* is the number of *measurements per coefficient*.

Following certain conditions, the normalized, per-bit Hamming distance (HD) d_h between the hashes of two vectors, obtained using (4.6), is bounded and depends on the precision parameter Δ. Additionally, d_h is correlated with the Euclidean distance (ED) d_e between the two original vectors if d_e is smaller than a certain threshold, also dependent on Δ. After this threshold, d_h converges quickly to its upper bound losing any relevant information on d_e. Conversely, it is provable that the HD between vectors further apart than this threshold provides no information whatsoever on the true distance that separates them and as such, this embedding provides information-theoretic security beyond the radius defined by this threshold, as long as the parameters A and w are kept secret from other parties [17, 78].

Secure modular hashing

A generalization of SBE, in SMH [76] the hash transformation Q_{SMH} is a random projection from $\mathbb{R}^N$ into $(\mathbb{Z}/k)^M$, where $\mathbb{Z}/k$ is the set of integers from 0 to $k-1$ and M is the number of hashes, such that [20]

$$Q_{SMH}(x) = \lfloor Ax + w \rfloor \ (mod\ k) \tag{4.7}$$

with $w \in \mathbb{R}^N \sim \text{unif}[0, k]$ and $A \in \mathbb{R}^{N \times M} \sim N\left(0, \frac{1}{\delta^2} I_N\right)$. The random matrix A and the random vector w are user-generated parameters that, as in SBE, must be treated as the framework's key and, therefore, must be kept secret in order to ensure its security.

4.4   PPML for paralinguistic and extralinguistic tasks

Having previously described the paralinguistic and extralinguistic tasks, on one hand, and the common cryptographic methods on the other, this section attempts to join both in order to give an overview of privacy-preserving paralinguistic and extralinguistic tasks. The first part of this section briefly reviews some works targeting tasks not related to health. The second part of this section describes in much more detail our own efforts for health-related tasks – the focus of the chapter – both using NNs and support vector machines (SVMs).

4.4.1   PPML for non-health-related tasks

To the best of our knowledge, the first contribution to privacy-preserving paralinguistics was made by Dias *et al.* [20]. In an approach similar to Jiménez *et al.* [76], the authors of this work applied SMH in combination with SVMs for privacy-preserving emotion recognition. In this work, as in Jiménez *et al.* [76], the authors take advantage of the characteristics and privacy guarantees of SMH and modify the radial basis function (RBF) kernel to work with HDs between SMH hashes instead of EDs between feature vectors. For this method to work, the SVM's training data has to be transformed using the same key (A, w) as the client's data, to ensure that the distances between the SVM's support vectors and the client's test vectors are meaningful. However, in this method, if the sever knows the key (A, w), there is nothing to stop it from training another model that uses the same features to obtain more information about the client, even if the data is protected under SMH. A solution for this would be to introduce a trusted third party. The server and the client would agree on a key to transform their data and the third party would be in charge of training the model, thus guaranteeing that no party has access to any information about the others' data. The server would then be able to use this model to test the client's data, without being able to use it to obtain out-of-domain information. Another possible solution would be for the client to generate the key, and apply it to the server's data using MPC methods.

Following the work of Gilad-Bachrach *et al.* [4, 58], Dias *et al.* [20] also provided a method for privacy-preserving emotion recognition by combining NNs and the BFV SHE cryptosystem to build an encrypted NN (ENN). In this method, all operations in the NN are replaced with their HE counterparts. Since BFV only allows multiplications and additions to be computed and activation functions are non-linear they cannot be computed using only these operations. Consequently, these functions need to be replaced by polynomial approximations in order to be computed with HE. Dias *et al.* [20] followed the approach of Chabanne *et al.* [58] and replaced activation functions with their Taylor series expansion at inference time. The authors of this work report an accuracy degradation of   2%–3% when comparing the private model to the baseline.

In an orthogonal approach, Nelus *et al.* [79–81] have proposed a series of works on gender discrimination while trying to hide the speaker's identity using Generative Adversarial Networks and Siamese NNs. In their work, the authors defined a feature extractor (computed locally), a gender classifier (computed remotely), and an attacker that has access to the same input as the gender classifier and who tries to identify the speaker. By training these networks adversarially, the authors were able to significantly drop the performance of the attacker. However, although the attacker's performance for this task is low, there is no guarantee that he/she cannot extract other sensitive characteristics from this representation as the attacker may be able to link these characteristics to the speaker's identity, breaking the speaker's privacy and putting him/her at risk. Additionally, by placing the feature extractor on the client's side, the server is forfeiting the privacy of part of its own model.

Thaine *et al.* [82] focused on the privacy-preserving extraction of low-level features. In particular, the authors proposed methods to extract bark frequency cepstral coefficients (BFCCs) and MFCCs from an encrypted signal using the BFV SHE scheme. The authors report that their method takes ~47 s to compute BFCCs from 100 frames of length 25. Moreover, the authors argue that it is inefficient to privately compute MFCC features, as it requires more expensive computations (such as the logarithm) to be performed. For this task, their method takes between 143 s and 346 s to compute the logarithm of a single encrypted value, depending on the logarithm's desired precision. The authors show that their approach introduces little to no accuracy degradation on the word error rate of an automatic speech recognition system trained with BFCCs and MFCCs extracted using their method. Although not directly applied to a paralinguistic or extralinguistic task, this method can be included as a first step in the pipeline of privacy-preserving methods for paralinguistic and extralinguistic tasks.

4.4.2   PPML for health-related tasks

The first two works related to privacy in health-related paralinguistic and extralinguistic tasks, to the best of our knowledge, were done by Teixeira *et al.* [21, 22], who applied variants of the ENN method of Dias *et al.* [20] to the detection and assessment of three speech-affecting disorders: depression, PD, and the common cold. Instead of using the approach of Chabanne *et al.* [58], Teixeira *et al.* followed the approach of Hesamifard *et al.* [57] and replaced activation functions with a Chebyshev polynomial approximation at both training and inference time. In their second work, *Teixeira et al.* [22] also took advantage of BFV's batching capabilities to compute several predictions at the same time, thus amortizing the effective cost of each prediction. To this end, the authors had to convert every weight in the network to integers. Furthermore, to avoid having to scale the network's inputs, these were quantized using μ-law quantization. The authors of both works state that their method yielded negligible accuracy degradation. It is nonetheless important to note that the results reported by these two works correspond to those obtained with the development set, which was used to tune the model's hyper-parameters. For this reason, these results may not reflect the true performance of these models on unseen test data. In terms of computational performance Teixeira *et al.* [22] report ~4.5 s for a single prediction without the use of batching, and ~23 s for 16 384 simultaneous predictions, yielding an amortized cost of ~1.4 ms per prediction.

4.4.3   Private SVM+RBF for health-related tasks

SVMs are powerful, and computationally light, discriminators that are able to perform well in a wide variety of tasks, including those where data are scarce – a frequent scenario in health-related paralinguistic and extralinguistic tasks. There has been a wide variety of works on privacy-preserving SVM inference, with several works implementing this classifier with both linear and polynomial kernels [83–87]. On the other hand, few works have proposed solutions for private SVM inference using the RBF kernel [88–90]. This can be justified by the combination of two factors: the RBF kernel requires the computation of the ED (cf. (4.8)) and, most importantly, the computation of the *exp*(.) function. While computationally heavy, the first can be solved through HE or secure MPC protocols. The second, however, is much harder, as it requires computing a polynomial approximation of the function:

$$d_E(x,y) = \|x_i - y_i\| = \sqrt{\sum_{i=0}^{M}(x_i - y_i)^2} \tag{4.8}$$

Bringer *et al.* [89] propose the use of a variation of the GSHADE [67] protocol to compute the RBF kernel. Nonetheless, the authors only report the computational performance of their method, and do not provide values for the results obtained. Some works have proposed de-centralized training and inference of SVMs with the RBF kernel, using set-intersection from data shared between multiple parties. However, notwithstanding its potential applications for hospitals and clinics, this setting is different from our own , as we consider a centralized service and model [88]. Makri *et al.* [90] used a random sampler to approximate the RBF kernel, allowing them to only perform linear computations privately. In a different approach, several works [17, 19, 20, 76] take advantage of distance-preserving hashing functions, such as SMH, that are able to mask data vectors while maintaining a proportionality, up to a certain threshold, between the ED and the HD (cf. (4.9)) of the original and hashed vectors, respectively:

$$d_H(x,y) = \frac{1}{M}\sum_{i=0}^{M} x_i \oplus y_i \tag{4.9}$$

Using this property, the RBF kernel can be adapted to work with the HD. Nonetheless, all these works rely on the security of the corresponding hashing techniques. Thus, while the original vectors may be protected, these approaches do not take into account the fact that as long as the service provider holds the users' hashed data vectors and the transformation parameters, it can train other models to extract out-of-domain information about the users, breaching their privacy.

In the remainder of this section we detail a privacy-preserving implementation of the RBF kernel which exploits the fact that the computational cost of the HD is much lower than that of the ED, making the former much more suited to be computed in a privacy-preserving setting. Therefore, instead of a method for privacy, we use SMH as a way to accelerate the privacy-preserving computation of the RBF kernel [91]. The fact that the HD only requires computing sums of XORs of single bits, while the ED is defined as the square root of a sum of squares, of floating point values, makes it much easier to compute the HD in a privacy-preserving setting than the ED. This approach is evaluated in two health related tasks: PD and OSA classification.

4.4.3.1 Private RBF computation

We can start by considering the original RBF kernel, defined as

$$k(x, x_i') = \exp(-\gamma d_E^2(x, x_i')) \tag{4.10}$$

where d_E^2 is the ED between x and x_i'. Following what was discussed above, we can transform our data using SMH and replace the ED with the HD.

Since the HD is a sum of XORs (4.9), we can efficiently compute this distance privately using secret sharing (cf. Section 4.3.3.2). It is not possible to compute an exponentiation directly with HE or secret sharing; consequently, we need to compute an approximation of the function. Considering we also need to multiply and square the HD before exponentiating, instead of computing all of these operations individually, we can combine the three into a single function to approximate. Moreover, to avoid having to perform multiplications using secret sharing, we can also move the normalizing factor $1/M$ out of (4.9) and incorporate it into our new function:

$$k(d_H') = \exp(-\tfrac{\gamma}{M^2} d_H'^2) \tag{4.11}$$

where d_H' is the non-normalized HD and M is the size of the vector. Several options exist on how to approximate the function, including a Taylor series expansion or computing an approximation using least squares. Unfortunately, for low degree polynomials, these approximations have small convergence intervals, and diverge quickly out of them. On the other hand, as shown by Hesamifard *et al.* [57], Chebyshev polynomials can approximate a function within a given interval, which is much more suited to our case. Since it is composed of real-valued coefficients, this polynomial can be evaluated efficiently using CKKS (cf. Section 4.3.1) [56].

4.4.3.2 Private SVM computation

To complete the full privacy-preserving SVM computation (cf. (4.12)), we need to multiply the output of the approximated RBF kernel with constants α_i, accumulate the results and add the intercept term w_0. Considering the fact that the server is in possession of the polynomial and α_i coefficients of the SVM, he/she can pre-multiply them, avoiding an extra level of multiplication. Additionally, dividing w_0 by the number of support vectors allows us to add it to the constant term of the polynomial, to avoid the extra addition. Finally, we need to compute the *sign*(.) function, which can be done using GCs.

The private SVM computation can thus be summarized with the following steps:

1. The server generates the SMH key, applies the transformation to his data and trains his model. The server then sends the key to the client, so that he/she can apply the same transformation to his data.
2. The two parties secret-share their data and jointly compute the HD between every pair of vectors.
3. The client encrypts his/her share and sends it back to the server, who re-builds the true value using HE.
4. The server is then able to compute the remaining of the kernel using the polynomial approximation, and subtracts a random integer from the result, to secretly share it with the client, sending the result back to him/her.
5. The client and the server jointly compute the *sign*(.) function using GCs, after which the client receives the server's share to obtain the final result of the classification.

4.4.3.3   Experimental setup

OSA detection

The corpus used for OSA detection is an extended version of the Portuguese sleep disorders (PSD) corpus (detailed description of the recording protocol and speech tasks can be found in Botelho *et al.* [27]). The corpus includes read and spontaneous speech recordings of 30 (21 male, 9 female) OSA patients and 30 (11 male, 19 female) control speakers. Each speaker recorded 12 items (1 small text, 10 sentences, and 1 image description). The total duration of the corpus is 2 hours 9 minutes. We partitioned the corpus in 4-s-long audio files using overlapping windows, with a shift of 2 s, resulting in 1 793 and 1 702 control and patient samples, respectively. Each sample is represented by a vector of 109 KB features, as proposed by Botelho *et al.* [27].

PD detection

The corpus used for PD detection corresponds to a subset of the New Spanish Parkinson's Disease Corpus, collected at the Universidad de Antioquia, Colombia [92], composed of read sentences. The corpus includes 50 PD patients and 50 controls. This subset of the corpus has a duration of 59 minutes. As with the PSD corpus, we partitioned each audio file in 4-s-long audio files using overlapping windows, with a shift of 2 s, if the utterance was larger than 4 s, resulting in 661 patients and

655 control samples. Each sample is represented by a 114-dimensional KB feature vector, as proposed by Pompili *et al.* [34].

4.4.3.4   Model training and parameters

We compare the performance of two models: SVM which refers to the baseline model, implemented without the privacy-preserving framework; and PP-SVM (privacy-preserving SVM) that refers to the privacy-preserving framework. The parameters for each SVM model and the SMH parameters were optimized through a grid-search. All models were trained using leave-one-speaker-out (LOSO) cross-validation. For OSA's baseline model, our best results were obtained using $C = 10$ and $\gamma = 0.001$. For the corresponding PP-SVM model we found that $C = 1$, $\gamma = 10$, $k = 4$, $\Delta = 1/\delta^2 = 0.031$, and $mpc = 32$ yielded the best results. For PD's baseline model, we used $C = 10$, $\gamma = 0.001$. For PD's PP-SVM model we used $C = 1\,000$, $\gamma = 0.01$, $k = 2$, $\Delta = 0.001$, and $mpc = 4$. All SVM models were implemented and trained using Python's *scikit-learn* SVC classifier [93]. The SMH transformation and the custom RBF kernel using the HD were also implemented in Python.

4.4.3.5   Private SVM implementation details

As stated above, our method requires three steps: HD computation between the client's input and the server's support vectors, evaluating the polynomial approximation of the kernel, and computing the *sign*(.) function. The first step can be efficiently computed using secret sharing. For this step we used ABY's [14] implementation of the Arithmetic and Boolean secret sharing protocols, as well as the corresponding conversions between them. At the end of this step both the server and the client hold a random-looking share of the HD between the client's input and the server's support vectors. For this step we took advantage of ABY's SIMD capabilities, and encoded 64 bits in each shared value, to speed up computations. The second step involves approximating the RBF kernel using a polynomial. Since we trained and tested our model using LOSO cross-validation, we computed a different polynomial approximation for each fold using the HDs between every pair of training data vectors. In this way, we emulated real-world conditions, where a service has fixed training and test sets. Our experiments showed that a fifth degree Chebyshev polynomial yielded the best trade-off between computational performance and accuracy. An example of a fifth degree Chebyshev polynomial approximation of function (4.11) can be observed in Figure 4.2. To perform this step we used SEAL's [94] implementation of CKKS [56], using a *polynomial modulus* of 8 192 and a *coefficient modulus* composed of two 60-bit-long and three 40-bit-long small primes. We took advantage of CKKS's batching capabilities to encode all HDs into fewer ciphertexts, thus reducing the communication and computational costs. To compute the final *sign*(.) function we used ABY's implementation of Yao's GCs [14] to perform a *greater than* operation. For both libraries we used the default parameters for 128-bit security.

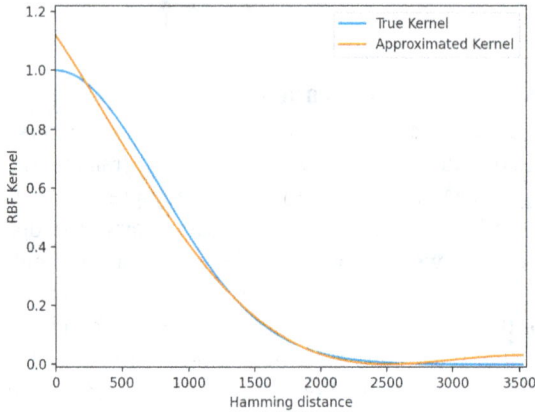

Figure 4.2   Fifth degree Chebyshev polynomial approximation of (4.11)

4.4.3.6   Classification results

The classification results for this task are presented in Table 4.2, in which the baseline corresponds to an SVM trained with data without any transformation, using the RBF kernel with ED; SVM + SMH corresponds to an SVM trained with SMH transformed data and using the RBF kernel with HD; and the Poly SVM+SMH corresponds to the results obtained training the SVM in the same way as in SVM+SMH, and performing inference over the test set with the polynomial approximation. We can see that the results obtained for OSA classification in the privacy-preserving framework are slightly worse than the baseline, but with a negligible difference (0.5% in terms of Unweighted Average Recall). On the other hand, the results achieved for PD classification surpassed the baseline, thus proving the validity of our approach.

4.4.3.7   Security and computational performance

The privacy of our method includes two components: the model and the client's input. Moreover, we need to evaluate the security of each component for every part of the protocol. Considering we used both HE and MPC, we evaluated the privacy of our method under each of these protocols. For the HE portion of our method, its security comes directly from the CKKS protocol [56], and neither party has access to the other's data. In the case of the MPC protocols, we can define our security under the *honest-but-curious* model, where both parties are expected to follow the protocol while trying to learn as much as possible about each other. We chose this assumption as it allows for more efficient models and is considered in the literature to be enough for most applications [6, 7, 18]. It is, nevertheless, possible to transition to the *malicious* model, at the cost of efficiency [15], as stated in Section 4.3.3.2. Assuming 109 features, 1432 support vectors, $k = 4$ and *mpc* = 32, resulting

Table 4.2   Results achieved for obstructive sleep apnea (OSA) and Parkinson's disease (PD) detection in terms of unweighted average precision, recall and F1 score

Method	OSA			PD		
	Precision (%)	Recall (%)	F1 score (%)	Precision (%)	Recall (%)	F1 score (%)
Baseline	69.0	68.9	68.9	78.6	78.6	78.6
SVM + SMH	68.3	68.2	68.2	80.1	79.7	79.6
Poly SVM + SMH	68.4	68.4	68.4	80.1	79.7	79.6

in hashed vectors of size 6 976, our implementation takes a total of ~600 ms for a single prediction (excluding communication time), using 3 MB of bandwidth. These results were obtained on a laptop with an Intel Core Quad-Core i5 CPU @ 1.40GHz and 16 GB of RAM.

4.5   Conclusions

The current state of the art in paralinguistic and extralinguistic health-related tasks indicates how mature this technology is becoming. However, ML models require a great deal of expertise and investment to be developed, making their distribution undesirable from a business point of view. Moreover, deep learning models are often computationally expensive, requiring hardware (i.e. Graphics Processing Units – GPUs) that may not be available for many users. Remote MLaaS solves these issues, but introduces ethical and legal concerns over patient privacy issues. The techniques described in Section 4.3 can be used to solve these concerns, but while they have been applied to numerous tasks in other fields, few contributions exist on the topic of privacy for paralinguistic and extralinguistic tasks, notwithstanding its relevance. This may be caused by several factors, including the difficulty in combining state-of-the-art ML methods with cryptographic primitives as well as the relative lack of maturity of the field. The fact that speech-based ML models have only recently started to obtain good results with health-related data *in-the-clear* explains why research on privacy techniques that are non-essential to obtain good speech-based disease classifiers has, so far, received little attention.

It is our belief that given the state of the art in health-related paralinguistic and extralinguistic tasks and current societal concerns on privacy, it is essential to increase the efforts to develop privacy-preserving techniques adapted to these tasks. Furthermore, as we have discussed in Section 4.2, linguistic cues can also be fundamental to the detection of certain diseases. While this adds strength to the need of protecting raw audio data, it also highlights the importance of the unexplored area of privacy-preserving methods for health based on transcribed text, making it a promising avenue for future work.

Acknowledgments

This work was supported by national funds through Fundação para a Ciência e Tecnologia (FCT) with reference UID/CEC/50021/2019 and grant BD2018 ULisboa.

References

[1]   European Parliament and Council. *On the protection of natural persons with regard to the processing of personal data and on the free movement of such data, and repealing Directive 95/46/EC (General Data Protection Regulation)*[Regulation 2016/679]; 2016 April.

[2] Nautsch A., Jasserand C., Kindt E., *et al.* 'The GDPR & speech data: reflections of legal and technology communities, first steps towards a common understanding'. *arXiv preprint.* 2019;1907.03458.

[3] Nautsch A., Jiménez A., Treiber A., *et al.* 'Preserving privacy in speaker and speech characterisation'. *Computer Speech & Language.* 2019;58:441–80.

[4] Gilad-Bachrach R., Dowlin N., Laine K., *et al.* 'CryptoNets: Applying neural networks to encrypted data with high throughput and accuracy'. ICML. vol. 48 of JMLR Workshop and Conference Proceedings; 2016. pp. 201–10.

[5] Liu J., Juuti M., Lu Y., *et al.* 'Oblivious neural network predictions via MiniONN transformations'. ACM SIGSAC Conference on Computer and Communications Security; 2017. pp. 619–31.

[6] Riazi M.S., Weinert C., Tkachenko O., *et al.* 'Chameleon: A hybrid secure computation framework for machine learning applications'. Asia Conference on Computer and Communications Security. ACM; 2018. pp. 707–21.

[7] Juvekar C., Vaikuntanathan V., Chandrakasan A. 'GAZELLE: A low latency framework for secure neural network inference'. USENIX Security Symposium; 2018. pp. 1651–69.

[8] Mohassel P., Rindal P. 'ABY3: A mixed protocol framework for machine learning'. ACM SIGSAC Conference on Computer and Communications Security; 2018. pp. 35–52.

[9] Mishra P., Lehmkuhl R., Srinivasan A., *et al.* 'DELPHI: A cryptographic inference service for neural networks'. 29th USENIX Security Symposium; 2020.

[10] LeCun Y., Bottou L., Bengio Y., *et al.* 'Gradient-based learning applied to document recognition'. *IEEE.* 1998;86(11):2278–324.

[11] Krizhevsky A. *Learning multiple layers of features from tiny images* [Master's thesis]. Department of Computer Science, University of Toronto; 2009.

[12] Halevi S., Shoup V. 'Algorithms in HElib' in Garay J.A., Gennaro R. (eds.). *Advances in Cryptology.* Berlin, Heidelberg: Springer Berlin Heidelberg; 2014. pp. 554–71.

[13] Laine K., Chen H., Player R. 'Simple encrypted arithmetic library - SEAL v2.3.1'. *Microsoft.* 2017.

[14] Demmler D., Schneider T., Zohner M. 'ABY - A framework for efficient mixed-protocol secure two-party computation'. NDSS; 2015.

[15] Damgård I., Keller M., Larraia E., *et al.* 'Practical covertly secure MPC for dishonest majority–or: breaking the SPDZ limits'. European Symposium on Research in Computer Security; Springer; 2013. pp. 1–18.

[16] Keller M. 'MP-SPDZ: A versatile framework for Multi-Party computation. Cryptology ePrint Archive'; 2020. Report 2020/521.

[17] Portêlo J., Abad A., Raj B., *et al.* 'Secure binary embeddings of front-end factor analysis for privacy preserving speaker verification'. Interspeech; 2013. pp. 2494–8.

[18] Treiber A., Nautsch A., Kolberg J., *et al.* 'Privacy-preserving PLDA speaker verification using outsourced secure computation'. *Speech Communication.* 2019;114:60–71.

[19] Portêlo J., Abad A., Raj B., *et al.* 'Privacy-preserving Query-by-Example speech search'. *In: ICASSP. IEEE*. 2015:1797–801.

[20] Dias M., Abad A., Trancoso I. 'Exploring Hashing and Cryptonet based approaches for privacy-preserving speech emotion recognition'. *ICASSP*; *IEEE*; 2018. pp. 2057–61.

[21] Teixeira F., Abad A., Trancoso I. 'Patient privacy in paralinguistic tasks'. Interspeech; 2018. pp. 3428–32.

[22] Teixeira F., Abad A., Trancoso I. 'Privacy-preserving paralinguistic tasks'. ICASSP; 2019. pp. 6575–9.

[23] Ringeval F., Schuller B., Valstar M., *et al.* 'AVEC 2019 Workshop and Challenge: State-of-mind, detecting depression with AI, and cross-cultural affect recognition'. 9th International Audio/Visual Emotion Challenge and Workshop; 2019. pp. 3–12.

[24] Cummins N., Epps J., Breakspear M., *et al.* 'An investigation of depressed speech detection: Features and normalization'. Interspeech; 2011.

[25] Morales M., Scherer S., Levitan R. 'A cross-modal review of indicators for depression detection systems'. *Fourth Workshop on Computational Linguistics and Clinical Psychology — From Linguistic Signal to Clinical Reality*. Vancouver, BC: Association for Computational Linguistics; 2017. pp. 1–12.

[26] Pompili A., Abad A., de Matos D.M., *et al.* 'Pragmatic aspects of discourse production for the automatic identification of Alzheimer's disease'. *IEEE Journal of Selected Topics in Signal Processing*. 2020:1–11.

[27] Botelho C., Trancoso I., Abad A. 'Speech as a biomarker for obstructive sleep apnea detection'. *ICASSP*; *IEEE*; 2019. pp. 5851–5.

[28] Kalia L.V., Lang A.E. 'Parkinson's disease. Current neurology and neuroscience reports'. 2015.

[29] Gomez-Vilda P., Londral A.R.M., Rodellar-Biarge V., *et al.* 'Monitoring amyotrophic lateral sclerosis by biomechanical modeling of speech production'. *Neurocomputing*. 2015;151:130–8.

[30] Skodda S., Schlegel U., Hoffmann R., Saft C. 'Impaired motor speech performance in Huntington's disease'. *Journal of Neural Transmission*. 2014;121(4):399–407.

[31] Kent R.D., Weismer G., Kent J.F., *et al.* 'Acoustic studies of dysarthric speech: methods, progress, and potential'. *Journal of Communication Disorders*. 1999;32(3):141–86.

[32] Rusz J., Cmejla R., Ruzickova H., Ruzicka E. 'Quantitative acoustic measurements for characterization of speech and voice disorders in early untreated Parkinson's disease'. *The Journal of the Acoustical Society of America*. 2011;129(1):350–67.

[33] Orozco-Arroyave J.R., Belalcazar-Bolanos E.A., Arias-Londoño J.D., *et al.* 'Characterization methods for the detection of multiple voice disorders: Neurological, functional, and laryngeal diseases'. *IEEE Journal of Biomedical and Health Informatics*. 2015;19(6):1820–8.

[34] Pompili A., Abad A., Romano P., *et al.* 'Automatic detection of parkinson's disease: An experimental analysis of common speech production tasks used for diagnosis'. International Conference on Text, Speech, and Dialogue; Springer; 2017. pp. 411–9.

[35] Arias-Vergara T., Vásquez-Correa J.C., Orozco-Arroyave J.R., *et al.* Parkinson's Disease Progression Assessment from Speech Using GMM-UBM. Interspeech; 2016. pp. 1933–7.

[36] Hauptman Y., Aloni-Lavi R., Lapidot I., *et al.* 'Identifying distinctive acoustic and spectral features in Parkinson's disease'. Proc. Interspeech 2019; 2019. pp. 2498–502.

[37] Laaridh I., Kheder W.B., Fredouille C., *et al.* 'Automatic prediction of speech evaluation metrics for dysarthric speech'. Interspeech; 2017. pp. 1834–8.

[38] Moro-Velazquez L., Villalba J., Dehak N. 'Using X-vectors to automatically detect Parkinson's disease from speech'. ICASSP; IEEE; 2020. pp. 1155–9.

[39] Perero-Codosero J.M., Espinoza-Cuadros F., Anton-Martin J., *et al.* 'Modeling obstructive sleep apnea voices using deep neural network embeddings and domain-adversarial training'. *IEEE Journal of Selected Topics in Signal Processing*. 2019;14(2):240–50.

[40] Zargarbashi S., Babaali B. 'A multi-modal feature embedding approach to diagnose Alzheimer disease from spoken language'. 2019.

[41] Du W., Morency L.P., Cohn J., *et al.* 'Bag-of-acoustic-words for mental health assessment: A deep autoencoding approach'. Interspeech; 2019. pp. 1428–32.

[42] Schmitt M., Janott C., Pandit V., *et al.* 'A bag-of-audio-words approach for snore sounds' excitation localisation'. Speech Communication; 12. ITG Symposium; VDE; 2016. pp. 1–5.

[43] Brasser F., Müller U., Dmitrienko A., *et al.* 'Software grand exposure: SGX cache attacks are practical'. 11th USENIX Workshop on Offensive Technologies (WOOT); 2017.

[44] Schwarz M., Weiser S., Gruss D. 'Practical enclave malware with Intel SGX'. International Conference on Detection of Intrusions and Malware, and Vulnerability Assessment; Springer; 2019. pp. 177–96.

[45] Murdock K., Oswald D., Garcia F.D., *et al.* 'Plundervolt: Software-based fault injection attacks against Intel SGX'. IEEE Symposium on Security and Privacy (SP); IEEE; 2020.

[46] Dwork C. 'Differential privacy'. *Encyclopedia of Cryptography and Security*. 2011:338–40.

[47] Konečný J., McMahan H.B., FX Y., *et al.* 'Federated learning: Strategies for improving communication efficiency'. NIPS Workshop on Private Multi-Party Machine Learning; 2016.

[48] Costan V., Devadas S. 'Intel SGX explained'. *IACR Cryptology ePrint Archive*. 2016;2016(086):1–118.

[49] Paillier P. 'Public-key cryptosystems based on composite degree residuosity classes. Advances in cryptology'. Advances in Cryptology. vol. 1592 of Lecture Notes in Computer Science; 1999. pp. 223–38.

[50]   Rivest R.L., Shamir A., Adleman L. 'A method for obtaining digital signatures and public-key cryptosystems'. *Communications of the ACM.* 1978;21(2):120–6.

[51]   Elgamal T. 'A public key Cryptosystem and a signature scheme based on discrete logarithms'. *IEEE Transactions on Information Theory.* 1985;31(4):469–72.

[52]   Gentry C. *A Fully Homomorphic Encryption Scheme.* Stanford University; 2009.

[53]   Chillotti I., Gama N., Georgieva M., *et al.* 'TFHE: fast fully homomorphic encryption over the torus'. *Journal of Cryptology.* 2020;33(1):34–91.

[54]   Brakerski Z., Gentry C., Vaikuntanathan V. '(Leveled) fully homomorphic encryption without bootstrapping'. *ACM Trans Comput Theory.* 2014;6(3):13:1–13:36.

[55]   Fan J., Vercauteren F. 'Somewhat practical fully homomorphic encryption'. *IACR Cryptology ePrint Archive.* 2012. 2012:144. Informal publication.

[56]   Cheon J.H., Kim A., Kim M., *et al.* 'Homomorphic encryption for arithmetic of approximate numbers'. International Conference on the Theory and Application of Cryptology and Information Security; Springer; 2017. pp. 409–37.

[57]   Hesamifard E., Takabi H., Ghasemi M. 'Cryptodl: Deep neural networks over encrypted data'. *arXiv preprint.* 2017;1711.05189.

[58]   Chabanne H., de Wargny A., Milgram J., *et al.* 'Privacy-preserving classification on deep neural network'. IACR Cryptology ePrint Archive; 2017. p. 35.

[59]   Chou T., Orlandi C. 'The simplest protocol for oblivious transfer'. *4th International Conference on Progress in Cryptology.* 2015;9230:40–58.

[60]   Schneider T., Zohner M. *GMW vs Yao? Efficient Secure Two-Party Computation with Low Depth Circuits. In: Financial Cryptography and Data Security.* Berlin Heidelberg: Springer; 2013. pp. 275–92.

[61]   Rabin M O. 'How to exchange secrets with oblivious transfer'. *IACR Cryptology ePrint Archive.* 2005;2005(01):187.

[62]   Beaver D. Precomputing oblivious transfer. Annual International Cryptology Conference; Springer; 1995. pp. 97–109.

[63]   Ishai Y., Kilian J., Nissim K. *Extending Oblivious Transfers Efficiently. In: Advances in Cryptology.* Berlin, Heidelberg: Springer; 2003. pp. 145–61.

[64]   Beimel A. *Secret-Sharing Schemes: A Survey. In: Coding and Cryptology.* Berlin, Heidelberg: Springer; 2011. pp. 11–46.

[65]   Snyder P2014Yao's Garbled circuits: recent directions and implementationsLiterature review, Dept of Computer Science, University of Illinois at Chicago

[66]   Yao A.C. 'How to generate and exchange secrets'. *27th Annual Symposium on Foundations of Computer Science (sfcs 1986)*; 1986. pp. 162–7.

[67]   Bringer J., Chabanne H., Favre M., *et al.* 'GSHADE: Faster privacy-preserving distance computation and biometric identification'. 2nd ACM workshop on Information hiding and multimedia security; 2014. pp. 187–98.

[68] Pinkas B., Schneider T., Smart N.P., *et al.* 'Secure two-party computation is practical'. International Conference on the Theory and Application of Cryptology and Information Security; Springer; 2009. pp. 250–67.

[69] Zahur S., Rosulek M., Evans D. 'Two halves make a whole: Reducing data transfer in Garbled circuits using half gates'. Cryptology ePrint Archive, Report 2014/756; 2014.

[70] Kolesnikov V., Schneider T. 'Improved garbled circuit: Free XOR gates and applications'. International Colloquium on Automata, Languages, and Programming; Springer; 2008. pp. 486–98.

[71] Goldreich O., Micali S., Wigderson A. 'How to play ANY mental game'. 19th Annual ACM Symposium on Theory of Computing. STOC '87; NewYork, NY, USA, ACM; 1987. pp. 218–29.

[72] Ben-Or M., Goldwasser S., Wigderson A. 'Completeness theorems for non-cryptographic fault-tolerant distributed computation'. 20th Annual ACM Symposium on Theory of Computing; ACM; 1988. pp. 1–10.

[73] Damgård I., Keller M., Larraia E., *et al.* 'Practical covertly secure MPC for dishonest majority–or: Breaking the SPDZ limits'. European Symposium on Research in Computer Security; Springer; 2013. pp. 1–18.

[74] Lindell Y. 'Secure multiparty computation (MPC)'. *IACR Cryptology ePrint Archive*. 2020;2020:300.

[75] Evans D., Kolesnikov V., Rosulek M. 'A pragmatic introduction to secure multi-party computation'. *Foundations and Trends® in Privacy and Security*. 2017;2(2-3).

[76] Jiménez A., Raj B., Portêlo J., *et al.* 'Secure modular hashing'. *WIFS*. IEEE; 2015. pp. 1–6.

[77] Portêlo J. 'Privacy-preserving frameworks for speech mining'. Instituto Superior Técnico; 2015.

[78] Boufounos P., Rane S. 'Secure binary embeddings for privacy preserving nearest neighbors'. WIFS; IEEE; 2011. pp. 1–6.

[79] Nelus A., Martin R. 'Gender discrimination versus speaker identification through privacy-aware adversarial feature extraction'. Speech Communication; 13th ITG-Symposium; 2018. pp. 1–5.

[80] Nelus A., Martin R. 'Privacy-aware Feature Extraction for Gender Discrimination versus Speaker Identification'. ICASSP 2019 - 2019 IEEE International Conference on Acoustics, Speech and Signal Processing (ICASSP); 2019. pp. 671–4.

[81] Nelus A., Rech S., Koppelmann T., *et al.* 'Privacy-preserving Siamese feature extraction for gender recognition versus speaker identification'. Interspeech; 2019. pp. 3705–9.

[82] Thaine P., Penn G. 'Extracting Mel-frequency and bark-frequency cepstral coefficients from encrypted signals'. Interspeech; 2019. pp. 3715–19.

[83] Laur S., Lipmaa H., Mielikäinen T. 'Cryptographically private support vector machines'. 12th ACM SIGKDD International Conference on Knowledge Discovery and Data Mining; ACM; 2006. pp. 618–24.

[84] Rahulamathavan Y., Phan R.C.W., Veluru S., *et al.* 'Privacy-preserving multi-class support vector machine for outsourcing the data classification in cloud'. *IEEE Transactions on Dependable and Secure Computing*. 2013;11(5):467–79.

[85] Bost R., Popa R.A., Tu S., *et al.* 'Machine learning classification over encrypted data'. NDSS. vol. 4324; 2015. p. 4325.

[86] Barnett A., Santokhi J., Simpson M., *et al.* 'Image classification using nonlinear support vector machines on encrypted data'. *IACR Cryptology ePrint Archive*. 2017;2017:857.

[87] Makri E., Rotaru D., Smart N.P., *et al.* 'PICS: private image classification with SVM'. *IACR Cryptology ePrint Archive*. 2017;2017:1190.

[88] Vaidya J., Yu H., Jiang X. 'Privacy-preserving SVM classification'. *Knowledge and Information Systems*. 2008;14(2):161–78.

[89] Bringer J., El Omri O., Morel C., *et al.* 'Boosting GSHADE capabilities: New applications and security in malicious setting'. 21st ACM on Symposium on Access Control Models and Technologies; 2016. pp. 203–14.

[90] Makri E., Rotaru D., Smart N.P., *et al.* 'EPIC: Efficient private image classification (or: learning from the masters)'. *Cryptographers' Track at the RSA Conference*. Springer; 2019. pp. 473–92.

[91] Jimenez A., Raj B. 'Privacy preserving distance computation using somewhat-trusted third parties'. ICASSP. IEEE; 2017. pp. 6399–403.

[92] Orozco-Arroyave J.R., Arias-Londoño J.D., Bonilla J.F.V., *et al.* 'New Spanish speech corpus database for the analysis of people suffering from Parkinson's disease'. LREC; 2014. pp. 342–7.

[93] Pedregosa F., Varoquaux G., Gramfort A., *et al.* 'Scikit-learn: machine learning in python'. *Journal of Machine Learning Research*. 2011;12:2825–30.

[94] Laine K., Microsoft S. *Microsoft Research*. Redmond, WA; 2020.

Chapter 5

Voice privacy in biometrics: speaker de-identification

Paula Lopez-Otero[1], Laura Docio-Fernandez[2], and Carmen García-Mateo[2]

Speech is becoming an important way of interaction with technologies such as intelligent cars, banking, and mobile phones. Some of these applications imply privacy and security issues: in e-health applications, privacy is important for the users; transmitting speech via the Internet can allow undesired users to impersonate us using voice conversion, speech synthesis technologies, etc. This creates the need to remove the identity of the speaker from the speech recordings. De-identification is a process by which a data custodian alters or removes an individual's identifying information from a dataset, making it harder for users of the data to determine the identity of the data subjects while allowing for data reuse. In the case of speech, it consists of removing the information about an individual's identity from the speech signal, but preserving other features of interest that are present in the signal such as the message and speaker state. This chapter presents the main research challenges for speaker de-identification. In addition, a comparison of state-of-the-art techniques is performed in a common experimental framework.

5.1 Introduction

De-identification is a process by which a data custodian alters or removes an individual's identifying information from a dataset, making it almost impossible for data users to determine the identity of the subject from which the data were extracted while allowing for data reuse and share [1]. The data can be of very different nature: text [2–4], images [5, 6], video [7, 8], and, of course, speech [9–11]. If the source data is speech, the de-identification process is called speaker de-identification.

[1]CITIC research center for Information and Communications Technologies - Universidade da Coruña, A Coruña, Spain
[2]atlanTTic research center for Telecommunication Technologies - Universidade de Vigo, Vigo, Spain

A wide range of applications that needs to process the speech signal can benefit from speaker de-identification. Two broad fields of application can be envisaged: privacy-preserving, voice-driven transactions, and privacy protection of sensitive information. Online voice-driven banking services, court witness recordings, and medical recordings are three examples of practical interest. Actually, the last example might arouse quite an interest among the scientific community when developing speech-based illness detection systems such as Parkinson's disease [12], Alzheimer's disease [13], bipolar disorder [14], or depression [15].

In summary, speaker de-identification aims to eliminate the information about a speaker's identity from his/her speech signal while preserving other information present in it. This implies that identifying information disappears, but other features remain, such as spoken message, speaker age and gender, personality traits and emotional state, or other characteristics related to physical or mental state. The output of the speaker de-identification process is a speech-like signal ready to be listened by a human or processed by a computer. Apart from the obvious objective of removing a speaker's identity, three different key issues for speaker de-identification systems were pointed out in [16]:

- Naturalness: de-identified utterances must sound natural and their content must be intelligible.
- Universality, i.e., it should work for any speaker.
- Reversibility, also known as speaker re-identification, implies the recovery of the original speech signal from the de-identified speech.

In some applications, used for public environment monitoring and automatic speech recognition (ASR) apps, reversibility might be unnecessary or even a threat to privacy. When reversibility is not met then the data can no longer be linked to any individual, and the process is called anonymization.

As in voice conversion (VC) and speech synthesis, quality is an important feature in speaker de-identification. De-identified utterances must sound natural and their content must be intelligible. Nevertheless, the importance of this requirement depends on the application: when the de-identified utterances are going to be processed by an automatic system, naturalness becomes less crucial than other features as long as it does not affect the performance of the data processing pipeline. Intelligibility is more relevant when an ASR system is used to extract automatic transcriptions of the de-identified utterances.

Speaker de-identification techniques should achieve universality: it should be possible to de-identify any speaker regardless of whether there is training data from this speaker that can be used to train conversion functions, i.e., the de-identification technique should be speaker-independent. The most common de-identification techniques are based on VC approaches that require source and target speakers' utterances to train the conversion parameters. As will be seen in Section 5.3, most VC approaches require a parallel corpus between source and target speakers, even though recent advances have achieved promising results with non-parallel corpora. Regardless of this fact, having training utterances of the speaker to be de-identified is an unrealistic scenario that is not practical for real-world applications. The work presented in Magariños *et al.* [11, 16] described speaker de-identification techniques that achieved universality by using ad-hoc or pretrained transformation functions, respectively. The de-identified utterances obtained by means of these techniques based on VC do not resemble a given target function. Nevertheless, differently to VC, this is not relevant in speaker de-identification since the main target is removing personally identifiable information, not achieving speech that sounds as a specific speaker.

The last desirable feature of speaker de-identification highlighted in Magariños *et al.* [16] is reversibility, also known as speaker re-identification. Being able to recover the original utterance from its de-identified version can be required in some applications where the aim is to protect the identity of the speaker for data transmission or storage, but a human listener prefers to receive the original utterance. Nevertheless, when speech is further processed by an automatic system, this is not necessary in general.

As pointed out in Ribaric *et al.* [17], some experts make a difference between de-identification and anonymization. De-identification is considered a reversible process, i.e., given a de-identified speech utterance, it is possible to recover the original speech if the required parameters are available. Anonymization is focused on hiding the personal information from the speech recording, but recovering the original identity is not possible or of interest. As done in Ribaric *et al.* [17], this chapter makes no distinction between de-identification and anonymization, but the reversible nature of the methods is discussed for each case.

This chapter first provides a discussion about how to evaluate the performance of speaker de-identification in Section 5.2. Next, Section 5.3 reviews the most common techniques used for speaker de-identification. Later, Section 5.4 performs a comparison of techniques in a common experimental framework, and the conclusions are summarized by the end of the chapter.

5.2 How to evaluate speaker de-identification?

Given that the literature on speaker de-identification is not very extensive, there is not a clear agreement on how to evaluate the quality of the resulting speech. The desired qualities of de-identified speech, namely naturalness and speaker identity removal, can be evaluated either by human listeners or using speech processing technologies. These evaluation methods can be considered as subjective and objective measures, respectively. The rest of this section describes the different approaches used in the literature to assess the quality of de-identified speech.

5.2.1 Subjective measures

Subjective measures for speaker de-identification assessment encompass those tests that are carried out by human beings that give their own opinion on the goodness of the evaluated systems. These measures are mostly borrowed from the text-to-speech literature because some of the items to evaluate are common to speaker de-identification, namely intelligibility and naturalness.

One of the most extended tests to evaluate naturalness is the mean opinion score (MOS). In this test, participants are asked to rate the naturalness of the de-identified speech on a five-point scale ranging from one (very unnatural) to five (very natural) [18]. It must be noted that naturalness is an underspecified concept with a subjective component, so before asking participants to perform this test, they must be instructed on how to evaluate naturalness [18]. MOS is commonly used in VC to evaluate the naturalness of the converted speech [19–23], and was used by Bahmaninezhad *et al.* [24] to evaluate speaker de-identification performance.

Another subjective measure borrowed from the speech synthesis field is the differential MOS (DMOS), where two speech samples are compared and rated by the participants on a five-point scale ranging from one (very dissimilar) to five (very similar). This measure is suitable for speaker de-identification since it allows the comparison between original and de-identified speech in terms of both naturalness and similarity. The use of DMOS is quite extended in the VC literature [22] and has also been used for speaker de-identification [11].

The VC community, however, does not completely agree on the use of DMOS for comparing speaker similarity. As stated by Wester *et al.* [20], comparing the similarity between two voices is an unusual task, and people, in general, are not trained to perform it. Hence, in VC challenges, same/different experiments are conducted, which consist in presenting two speech samples to the participants and asking them to rate if they think the two speech samples could have been produced by the same speaker on a four-level scale ("same: absolutely sure," "same: not sure," "different: not sure," "different: absolutely sure") [20, 21].

XAB preference test is another subjective measure that is frequent in VC literature [25–27]. In this test, a triplet of speech samples X, A and B are presented, and the participant is asked to decide which sample, A or B, is more similar to sample X in terms of speaker individuality. The XAB test has been used in the speaker de-identification field to assess re-identification performance [11]. There also exists a similar test, namely the ABX test, which is commonly used to compare two different VC methods [19, 26, 28, 29]. In this test, two samples A and B are presented, and the participant is asked to decide whether a third sample X was generated by A or B.

All the aforementioned subjective measures are suitable for evaluating the naturalness of de-identified speech, but intelligibility is an important quality that must be taken into account. However, the amount of works in the literature focusing on intelligibility are not so wide, except for those related to speech disorders such as degenerative diseases that affect speech [30], electrolaryngees [31], body-conducted unvoiced speech enhancement [32] or silent speech interfaces [33], to cite some. In this literature (and that related to synthetic and converted speech in general), the use of MOS to assess intelligibility is common [23, 31]. There is an alternative to evaluate intelligibility in a more precise manner, which consists in asking a group of participants to transcribe a set of spoken utterances and then compute the word error rate (WER) [30, 32, 34–36]. For this purpose, sentences are sometimes generated following the system usability scale (SUS) test [37], which consists in generating semantically unpredictable sentences. This evaluation strategy based on the WER was used to assess speaker de-identification intelligibility in [38].

5.2.2 Objective measures

The subjective measures described in Section 5.2.1 are widely used in speech synthesis and VC literature, but they have some drawbacks that are causing them to be replaced by other alternatives mostly based on automatic speech processing strategies. Many applications of speaker de-identification aim at providing privacy protection when transmitting data through the Internet, and potential attackers would most likely make use of objective measures to perform their attacks. In addition, big data applications managing speech contents employ automatic approaches to process the data. Hence, achieving good speaker de-identification results according to objective measures is important in this field.

As happened with the subjective measures, there are objective measures for speaker de-identification that were borrowed from the speech synthesis and VC fields. This is a case of Mel-cepstral distortion (MCD), an error measure used in VC to compare transformed and reference utterances recorded by the target speaker [19, 27, 28, 39]. This measure has been used by Magariños *et al.* [11] to assess speaker de-identification and re-identification: in the former, MCD is expected to be high (i.e., the original utterance is expected to be dissimilar to the de-identified one), whereas in the latter MCD is ideally small (i.e., the re-identified utterance is similar to the original utterance).

As mentioned in Wester *et al.* [20], the ability of human beings to compare voices in terms of similarity is not quite developed. Hence, in the speaker de-identification community, the use of automatic speaker recognition approaches is becoming very popular for assessing the similarity between original and de-identified speech utterances. This evaluation strategy is easier to use for researchers since it does not require the collaboration of human listeners that are usually difficult to recruit. Therefore, most of the speaker de-identification literature makes use of either speaker identification [9–11, 16, 40, 41] or speaker verification [16, 24, 42–44] to evaluate speaker de-identification accuracy. The use of speaker verification is more common in the recent literature since the evaluation corpora usually have few speakers and, therefore, speaker identification approaches introduce a bias in the evaluation procedure (the probability of randomly choosing the right speaker is greater when the number of speakers is smaller).

As described in Section 5.2.1, the evaluation of intelligibility in a subjective manner is carried out by asking participants to transcribe speech utterances. The listeners are usually recruited by asking colleagues to perform the tests, but also through online platforms such as Amazon Mechanical Turk (MTurk).[1] However, the study presented in [34] reflected some disagreement on the intelligibility results obtained from MTurk

---

[1] www.mturk.com

listeners and others recruited in the authors' laboratory. These authors even carried out additional analyses to try to find the source of disagreement (background noise, headphone type, hearing ability, among others) [45]. Therefore, using ASR strategies to evaluate intelligibility seems reasonable. This method has been used to assess the intelligibility of articulatory-to-speech synthesis [46] and VC for laryngeal voices [47].

5.3   Speaker de-identification techniques

Both linguistic and non-linguistic information present in the speech signal are linked to speaker identity [48]. Linguistic characteristics refer mainly to language or dialect and to the use of certain words, whereas non-linguistic information, related to sociological factors or physiological attributes of the speaker, is more clearly linked to speaker individuality. Perceptually, the most important acoustic features characterizing speaker individuality include the third and the fourth formants, the fundamental frequency (F0) and the closing phase of the glottal wave.

Voice disguise can be used to hide the speaker's identity and therefore as a method of speaker de-identification. Two types of intentional voice disguises are differentiated in Wu *et al.* [49] and Perrot *et al.* [50]: in one of them, the modification of the voice is done by non-electronic means; and in the other, the modification is done using electronic devices. Non-electronic methods include using falsetto, pinching nostrils, speaking with object in mouth, among others, and electronic methods use electronic devices to alter the voice that, as we will see later, are very often found also in VC techniques. For example, Wu *et al.* [49] used the increased or lower voice pitch as a voice disguising technique, for which a scaling factor of pitch semitones is defined as the disguising factor.

However, the most extended approach for speaker de-identification consists of applying VC techniques in order to modify the voice characteristics of a source speaker into those of a target speaker, to resemble the target speaker voice. VC can be described as a regression problem for estimating a mapping or transformation function between the source and target features. This transformation modifies the physical characteristics of the voice without modifying the content of the message. Thus, an important aspect in VC systems is the choice of the speaker identity-related features to be converted. From a perceptual point of view, speaker identity information is extracted from the speech signal through both low- and high-level features simultaneously. High-level features refer mainly to speaking style and are related primarily to linguistic characteristics, such as language or dialect, the use of certain words, and other non-speech speaker-specific traits. Low-level characteristics refer to how the speaker's voice sounds and are related to voice source (average pitch frequency, pitch contour, pitch frequency fluctuation, glottal wave shape, among others) and vocal tract characteristics (spectral envelope, spectral tilt, format frequencies and bandwidths, to cite some examples) . Of all these characteristics, most state-of-the-art VC systems aim to modify the spectral envelope and the pitch.

As mentioned by Mohammadi and Kain [51], VC techniques can be classified in different ways depending on various factors. One factor is whether or not a parallel corpus is required in the training phase. This parallel corpus consists in recording utterances with the same linguistic content and where only the identity of the speaker [52] varies. A second factor is whether they are text-dependent or text-independent [53]. Text-dependent techniques need word- or phone-level transcriptions of recorded utterances from both source and target speakers. For text-independent approaches there is no transcription available, so speech segments of similar linguistic content need to be found before training the conversion function [54]. A third factor is the language of source and target speakers, leading to language-dependent or language-independent techniques. In language-independent or cross-language techniques, source and target speakers speak different languages [55, 56], which can cause mapping problems due to phonetic differences between both languages, as there is not always a one-to-one correspondence between all phonemes. This problem can be solved by using a combination of non-parallel and text-independent approaches.

In general, three key aspects have to be considered in the development of a VC system: the conversion of spectral parameters, the conversion of prosodic parameters and the generation of the waveform.

According to the literature, most of the proposed VC systems are based on a spectral feature mapping using parallel data. These systems first convert the speech features extracted from the source speaker's voice to those of the target speaker using a mapping function that must be trained, and then the converted speech features are used to generate a converted speech waveform using high-quality speech analysis/synthesis techniques. In addition, before training the mapping function the voice signals between source and target speakers must be time aligned frame by frame using, for example, a dynamic time warping (DTW) approach. The main drawback of these systems is that they need a huge amount of parallel training data for obtaining robust and high-quality VC. However, such parallel data are difficult to obtain and are not always available. To overcome this difficulty, significant research effort is recently being devoted to approaches that do not require parallel data.

Over the years, many VC techniques based on statistical methods for spectral feature mapping have been proposed. The most common techniques are briefly reviewed below. Table 5.1 gives a summary of the performance of some of these VC methods.

5.3.1 Codebook mapping

These methods are based on finding the correspondence between the vectors of two codebooks, i.e., of the source and target speakers, which determines the mapping between both speakers. A conversion system was proposed in [57] using this technique. In the training stage, a set of utterances is used to create the codebooks of source and target speakers by applying vector quantization (VQ) at the frame level. Then, DTW is applied to find correspondences between the vectors of the two codebooks that correspond to the same words (alignment). These correspondences are accumulated as histograms, which can be used as a weighting function and to define the mapping codebook as a linear combination of the target vectors. This process

Table 5.1   Overview of some voice conversion techniques

VC technique	References	Database	Metric	Performance
Codebook mapping	[57, 58]	TIMIT	ABX, ABX	60.75%, 76.4%
Gaussian mixture model	[25, 59]	CMU ARCTIC	MCD (dB), MCD (dB)	8.2, 4.11
Frequency warping	[60, 61]	TC-STAR (Spanish), CMU ARCTIC	mcd[a], MCD (dB)	1.4, 5.50
Deep learning	[19, 62, 63]	CMU ARCTIC, VCC2016	MCD (dB), MCD (dB)	5.77, 6.08

[a]mean cepstral distance
ABX, ABX Test;MCD, Mel-cepstral distortion;mcd, mean cepstral distance.

is repeated iteratively to refine the codebook mapping. In the conversion stage, the source speaker characteristic vectors are first quantified based on his/her codebook, and then the converted feature vectors are determined by the mapping codebook. This codebook mapping method can convert the speaker's identity quite success-fully. However, since in conventional VQ only one of the vectors is used to represent the correspondence, frame by frame discontinuities occur, which degrade the quality of the generated speech. The use of fuzzy VQ is proposed in [64] to improve the quality of converted speech, and a weighted average of codebooks is used in [58, 65] to represent each speech frame in order to generate a smooth transition through successive frames.

5.3.2   Gaussian mixture model

In order to cope up with the problem of codebook mapping, Gaussian mixture model (GMM)-based methods were proposed for soft clustering and continuous mapping of spectral features [59, 66]. The GMM partitions the speaker's feature space into a finite number of overlapping classes that allows to learn a linear transformation for each class. From there, the mapping or conversion function is defined as a statisti-cally weighted combination of these linear functions. In these methods, the spectral features of the source speaker are also converted frame by frame independently. Therefore, they do not consider the continuous nature of spectral features, which may result in the discontinuity of feature trajectories, and thus degrade the speech signal quality. To refine the estimated feature trajectories, modeling the spectral fea-tures with dynamic components was proposed in [25]. Specifically, a maximum like-lihood parameter-generation method was adopted to generate the converted spec-tral feature trajectories. By using dynamic features, temporal information between frames is considered in feature prediction, leading to smooth feature trajectories and improving the quality of converted speech.

Although GMM-based methods can generate speech highly similar to the target speaker, they suffer from oversmoothing [25] problems due to using the minimum mean square error, or the maximum likelihood function, as optimization criteria. This results in a statistical averaging causing the converted speech not to capture well the desired details of temporal and spectral dynamics. In addition, these methods generally employ low-resolution and low-dimensional spectral features such as Mel-cepstral coefficients (MCEPs) or line spectral frequencies, which fail to model spectral details and long-term information. These two aspects, statistical averaging and low-resolution features, cause most GMM-based methods to generate muffled speech [25, 60]. To enhance the temporal variation and alleviate the oversmoothing problem, global variance of feature trajectories [25] and modulation spectrum [67] were proposed.

5.3.3   Frequency warping

Frequency warping (FW) approaches for VC try to overcome the oversmoothing problem of GMM-based VC. These techniques aim to find an optimal warping function that shifts the frequency axis of the source spectrum towards the target spectrum. Different types of FW-based approaches have been proposed in the literature. Early approaches are somehow based on vocal tract length normalization (VTLN), which aims at compensating the effect of speaker-dependent vocal tract lengths by warping the frequency axis of the power spectrum [68]. Specifically, different warping functions have been proposed and studied, from warping functions realized by a single parameter [55, 69] to functions represented as a piecewise linear function [60, 61, 70]. Nevertheless, using a single parameter FW function may not adequately describe the mapping of spectral details, so the most widespread approaches use piecewise linear functions, which cause the whole frequency axis to be warped in the same direction, either to lower or to higher frequencies.

Several parametric FW functions were studied by Sündermann and Ney [55], including piecewise linear, power, quadratic, and bilinear functions, which achieved high-quality speech even though the speaker identity is not fully converted. In that work the acoustic feature space is divided into artificial phonetic classes, which are used to estimate the parameters of class-dependent VTLN warping functions. Later, in [61] an iterative algorithm was proposed to estimate the VTLN parameters. These approaches lead to the discontinuity of the converted spectra across classes, so, in order to diminish this effect, a weighted FW function that uses a piecewise linear FW is estimated from the formant frequencies [60]. Each acoustic class is modeled by a GMM, and an optimal FW function is assigned to each of the acoustic classes. In the conversion stage, the warping function for each frame is obtained by combining the set of trained functions according to the probabilities of a frame belonging to each of the acoustic classes.

As the warping process does not remove any spectral details, the quality of converted speech is high. However, FW itself only transforms the frequency axis without considering the amplitude of the spectrum. Hence, such approaches suffer from a speaker similarity problem. To address this, amplitude scaling (AS) [60, 61, 69–71], also known as energy correction filter, is proposed to compensate for the amplitude difference between the warped and the target spectra.

5.3.4   Deep learning techniques

The power and the success of deep learning techniques in the field of machine learning in the recent past have led many researchers to also apply these techniques to the VC problem. The use of approaches based on deep neural networks (DNNs) allows non-linear mapping between source and target features, with little restrictions on the type and dimension of the features to be modeled and transformed.

A very early work on the use of neural networks for spectral conversion, focused on formant transformation, was presented in [72], where a feedforward network with two hidden layers was used. Some other pioneering works focused on DNN-based spectral conversion showed that different architectures of feedforward DNNs are able to convert the spectral features effectively [19, 62]. Specifically, the work proposed by Desai *et al.* [19] investigated different architectures of feedforward networks for MCEPs feature mapping. Restricted Boltzmann machines (RBMs) and their variations were also proposed for spectral feature mapping by Lee [29, 73]. For example, an RBM can be used to model the joint distributions of the source and target spectral features as in [74], while a deep belief network is proposed in [73] to build high-order eigenspaces of the source and target speakers, where it is easier to convert the source speech into the target speech than in the cepstrum space.

Recently, variational autoencoders (VAEs) were also proposed for VC [63, 75]. VAEs are probabilistic generative autoencoders that can infer latent codes from data and generate new data from them by jointly learning inference and generative networks. This allows them to model speech attributes such as speaker identity and linguistic content [76]. However, in these DNNs the spectral features are modeled independently, ignoring the temporal dependencies in speech signals. Recurrent neural networks (RNNs) were proposed to try to solve it. These networks implicitly model temporal behavior by using recurrent connections that help retain information over a long period of time. Even though standard RNNs are able to capture the temporal information among speech frames, they have limited capabilities when modeling long-range context dependencies. Bidirectional long- short-term memory-based RNNs were proposed to overcome this problem; they can learn long-range contextual information in both forward and backward directions. By taking bidirectional history into consideration, these networks outperform traditional DNN-based VC techniques by improving the naturalness and continuity of the converted utterances.

5.4   Experiment definition

This section aims to evaluate the performance of different speaker de-identification approaches on two different settings: speaker-dependent, where a specific transform is trained for each speaker, and speaker-independent, where the transform applied

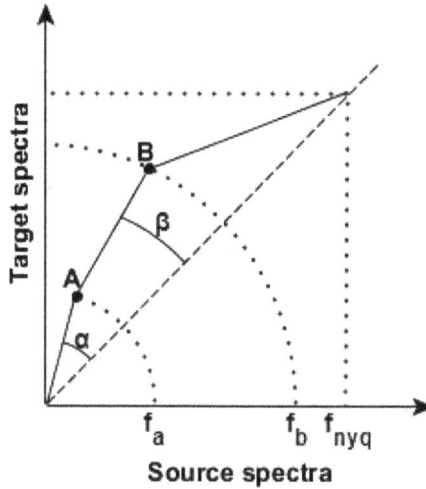

Figure 5.1   Piecewise linear approximation of a FW

for de-identification is independent of the input speaker, i.e., it was not trained specifically for that speaker.

First, speaker de-identification methods were chosen for these experiments, which make use of the most common VC techniques available in the literature. Specifically, two techniques based on FW + AS and two based on deep learning approaches were selected for these experiments.

5.4.1   Piecewise definition of transformation functions

This technique, proposed by Magariños *et al.* [11], performs speaker de-identification by VC in a non-strict manner: in VC, a source speaker is converted to a target speaker but, in this strategy, a source speaker is converted to an unknown speaker by manually defining the conversion parameters. One of the advantages of this method is that a training corpus is not necessary since the converted parameters are computed following predefined functions.

This approach applies a linear transform in the cepstral domain based on FW combined with AS [68]:

$$y = Ax + b \tag{5.1}$$

where x is a Mel-cepstral vector, A represents a FW matrix, b is an AS vector, and y is the transformed version of x.

A training stage is typically performed in FW + AS strategies in order to obtain the parameters of A and b. In this method, instead of training the parameters of A and b, they are manually defined following some guidelines, as depicted in Figure 5.1. First, the FW curve is simplified and defined piecewise using three linear functions: the discontinuities of the curve are set at frequencies f_a and f_b; α is the angle between the 45° line and the first linear function; and β is the angle between the 45° line and the second linear function, defined as $\beta = k\alpha$ $(0 < 1)$. When α is greater than 0, formants are moved to higher frequencies, resulting in a male-to-female transformation function. On the contrary, negative values of α lead to female-to-male conversion functions.

The AS vector **b** is defined by randomly giving values to a set of weighted Hanning-like bands equally spaced in the Mel-frequency scale [77] as fully described by Magariños *et al.* [11]. Finally, FW + AS is complemented with a scaling of the F0 proportional to the value of α, since it showed to dramatically improve de-identification performance [11].

In this system, Ahocoder [78] was used to extract 40 Mel-cepstral coefficients (MCEPs), F0, and band aperiodicity (BAP) features. Given an utterance to de-identify, its MCEPs are converted applying the FW and AS parameters described above, and the F0 is scaled. Then, Ahocoder is used to synthesize the de-identified speech utterance using these features and the BAP features, which remain unchanged. The free parameters of the system f_a, f_b, and k were set to 700 Hz, 3 000 Hz, and 0.5 according to Magariños *et al.* [11].

Given that this de-identification strategy uses a specific conversion function according to the gender of the speaker, the gender of the speaker must be identified before de-identification. Hence, a gender detector based on GMM log-likelihood ratio was built: given male and female GMMs, the likelihood of a test utterance for each GMM is computed and their log-likelihood ratio is calculated, assigning the most likely gender to the utterance [79]. The GMMs were obtained by training with gender-dependent data included in the FA subcorpus of the Albayzin database [80], which comprises around 4 hours of speech uttered by 200 different speakers (100 of either gender). The number of mixtures of the GMMs was set to 1 024 according to Lopez-Otero *et al.* [44], and the features used were 19 Mel-frequency cepstral coefficients (MFCCs) augmented with energy, delta, and acceleration coefficients.

The dynamic FW technique was used to obtain the FW parameters [61], whereas the AS vector is computed as the cepstral difference between the averaged target vectors $\bar{y}$ and the warped version of the averaged source vectors $\bar{x}$:

$$b = \bar{y} - A\bar{x} \tag{5.2}$$

As shown in [11], performing adaptation of the F0 on top of FW + AS transformation dramatically improves de-identification performance. Hence, in this work, a mean and variance adaptation of the F0 is applied:

$$\log \hat{f_0^t} = \frac{\sigma_t}{\sigma_s}(\log f_0^s - \mu_s) + \mu_t \tag{5.3}$$

where f_0^s represents the frame-level values of the source speaker's F0; μ_s, σ_s, μ_t, σ_t are the mean and standard deviations of the source and target F0 in the log domain, respectively; and $\hat{f_0^t}$ is the adapted F0 of the target speaker.

5.4.2   *Pretrained transformation functions*

The technique described by Magariños *et al.* [16] was used in this work for speaker-independent de-identification. In this strategy, first a set of conversion functions is

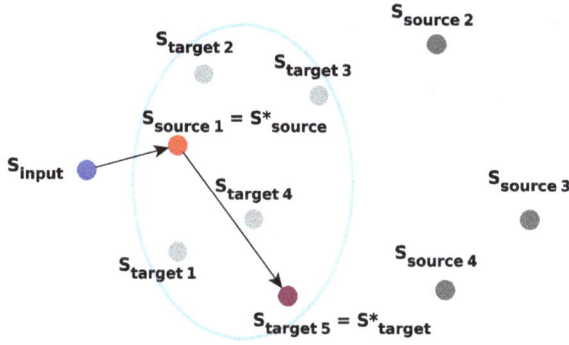

Figure 5.2 *Diagram of the proposed speaker selection strategy for voice conversion. The dots represent the different speakers, and those inside the blue ellipse are training speakers that share a parallel corpus*

trained given a set of source and target speakers that share parallel corpora among them. Then, given an input utterance to be de-identified, the source speaker that is most similar to the input speaker is chosen as *source*, and the target speaker that is most dissimilar to *source* is selected as *target*. Therefore, the transformation function previously learned for the chosen *source* and *target* speakers is the one applied for de-identification of the input speech. This leads to speech that does not sound exactly as the target speaker but the identity of the input speaker is removed, which is the aim of speaker de-identification [16].

A method for selecting the source and target speakers is necessary. The procedure considered in this work for this purpose is depicted in Figure 5.2. Given an input speaker S_{input} and a set of n_s potential source speakers from the parallel corpus $S_{source} = \{S_{source_1}, \ldots, S_{source_{n_s}}\}$, first the selected source speaker S^*_{source} for the transformation function is the one from S_{source} that maximizes the similarity with S_{input}. Once S^*_{source} is selected, given a set of n_t potential target speakers $S_{target} = \{S_{target_1}, \ldots, S_{target_{n_t}}\}$, the selected target speaker S^*_{target} is the one from S_{target} that minimizes the similarity with S^*_{source}. The motivation behind this method lies in the premise that the objective of de-identification is not mimicking a target speaker but producing speech in which the identity of the speaker is not recognizable. Hence, choosing a source speaker that is similar to the input speaker ensures that the transformation parameters are as suitable as possible for the input speaker. In addition, selecting the most dissimilar target speaker maximizes the chance of achieving speech that sounds very different to the input speaker. The main advantage of this method is the possibility of de-identifying speech from any speaker regardless of the availability of a parallel corpus including data from this input speaker.

In this work, the similarity between speakers is obtained using the i-vector paradigm [81] combined with probabilistic linear discriminant analysis (PLDA) scoring

[82]. This technique defines a low-dimensional space, namely total variability space, in which speech segments are represented by a vector of total factors, commonly known as i-vector [81]. Given a speech utterance, its corresponding GMM supervector $\mathbf{M}$ can be decomposed as:

$$\mathbf{M} = \mathbf{m} + \mathbf{Tw} \tag{5.4}$$

where $\mathbf{m}$ is the speaker and channel-independent supervector, $\mathbf{T}$ is a low-rank total variability matrix, and $\mathbf{w}$ is the i-vector corresponding to the GMM supervector. A pair of i-vectors can be compared by computing their PLDA scoring [82].

Following the notation of the rest of this section, given an input speaker S_{input}, its i-vector $\mathbf{w}_{input}$ is extracted and compared with all the potential source speakers, selecting the one that maximizes the following:

$$S^*_{source} =_{i \in 1,...,n_s} \text{score}(\mathbf{w}_{input}, \mathbf{w}_{source_i}) \tag{5.5}$$

Equivalently, the target speaker is selected by minimizing the PLDA score between the source speaker and the potential target speakers:

$$S^*_{target} = \arg \min_{i \in 1,...,n_t} \text{score}(\mathbf{w}^*_{source}, \mathbf{w}_{target_i}) \tag{5.6}$$

In this system, the feature configuration and vocoder used in the piecewise strategy described in Section 5.4.1 were adopted. In addition, given that the selection of the most similar and dissimilar speakers relies on the i-vector paradigm, a universal background model (UBM) and a total variability matrix were trained using the DS1 partition of the Biosecure database [83], which consists of 13 hours of audio of 316 different speakers. The UBM, which is a GMM with diagonal covariance matrix, has 1 024 mixtures, and the dimension of the i-vectors was set to 100 according to Magariños *et al.* [16].

5.4.3   De-identification based on DNNs

This system uses a feedforward neural network to convert a speech frame $\mathbf{x}_t$ of a source speaker into another $\mathbf{y}_t$ that sounds as uttered by a target speaker:

$$\mathbf{h}_t = \mathcal{H}(\mathbf{W}^{xh}\mathbf{x}_t + \mathbf{b}^h) \tag{5.7}$$
$$\mathbf{y}_t = \mathcal{H}(\mathbf{W}^{xy}\mathbf{h}_t + \mathbf{b}^y) \tag{5.8}$$

where $\mathcal{H}(.)$ is a non-linear activation function in a hidden layer, $\mathbf{W}^{xh}$ and $\mathbf{W}^{xy}$ are the weight matrices, and $\mathbf{b}^h$ and $\mathbf{b}^y$ are the bias vectors [84].

This system is implemented in Merlin toolkit[2] [84]. In the training stage, features are extracted from the training utterances of the source and target speakers using WORLD vocoder [85] and the Speech Signal Processing Toolkit (SPTK) [86]. Specifically, 60 Mel generalized cepstral (MGC) coefficients, F0, and BAP features are obtained. The standard configuration of Merlin was modified so that

---

[2]https://github.com/CSTR-Edinburgh/merlin

BAP features remained unchanged since this yielded better performance in these experiments. After feature extraction, the feature vectors of the source and target training utterances are aligned by means of the DTW algorithm. The architecture of the feedforward network is as follows: 4 hidden layers of dimension 512 with *tanh* activation functions. A stochastic gradient descent optimizer was used in the training stage.

Given an utterance to de-identify, feature extraction is performed and its MGC and F0 features are converted following (5.8). Finally, the converted features, along with their corresponding BAP features, are synthesized using WORLD vocoder.

5.4.4 De-identification based on generative adversarial networks

This system was proposed by Saito *et al.* [87] for both speech synthesis and VC, and is used in these experiments for speaker de-identification. In this approach, generative adversarial networks (GANs) are used to avoid the oversmoothing effect observed in other DNN-based strategies. GANs combine two different neural networks: the discriminator tries to distinguish natural and generated samples, whereas the generator aims at deceiving the discriminator.

As in Section 5.4.3, given a set of T samples x_t of the training speaker, they are transformed into samples y_t that sound as spoken by the target speaker. This is done by iteratively training the generator and discriminator DNNs using a stochastic gradient descent optimizer. First, the loss of the discriminator is computed as

$$L_D^{(GAN)}(\mathbf{x}, \mathbf{y}) = -\frac{1}{T} \sum_{t=1}^{T} log \frac{1}{1+exp(-D(\mathbf{x}_t))} - \frac{1}{T} \sum_{t=1}^{T} log(1 - \frac{1}{1+exp(-D(\mathbf{y}_t))}) \quad (5.9)$$

Then, the parameters of the discriminator are updated and then the adversarial loss of the generator is computed as

$$L_{ADV}^{(GAN)}(\mathbf{y}) = -\frac{1}{T} \sum_{t=1}^{T} log \frac{1}{1+exp(-D(\mathbf{y}_t))} \quad (5.10)$$

The parameters of the generator are subsequently updated. The loss function used for training the GAN is

$$L_G(\mathbf{x}, \mathbf{y}) = L_{MGE}(\mathbf{x}, \mathbf{y}) + \omega_D \frac{E_{L_{MGE}}}{E_{L_{ADV}}} L_{ADV}^{(GAN)}(\mathbf{y}) \quad (5.11)$$

where

$$L_{MGE}(\mathbf{x}, \mathbf{y}) = \frac{1}{T}(\mathbf{y} - \mathbf{x})^T(\mathbf{y} - \mathbf{x}) \quad (5.12)$$

is the loss of the minimum generator error trainer as described in [87] and ω_D is a hyper-parameter that controls the weight given to the loss function of the GAN network.

In this system, speech utterances are represented using 60 MGCs, F0, and BAP features extracted with WORLD [85] and SPTK [86]. The training procedure described above is used to transform the MGC features, whereas the F0 is scaled as described in Section 5.4.2 and BAP features remain unchanged. In these experiments, ω_D was set to 1.

5.5   Evaluation corpora

Two corpora were used to evaluate the performance of the speaker de-identification approaches described in Section 5.4. One of them includes many speakers with few training data, whereas the other includes few speakers with more training data. The purpose of using these two corpora is evaluating the performance of those speaker de-identification techniques according to the amount of training data, and also assessing the relevance of having more or less speakers available as source speakers for speaker-independent de-identification.

The first dataset is the Voice Cloning Toolkit (VCTK) Corpus [88]. VCTK Corpus was designed for speech synthesis applications, but its characteristics make it suitable for VC (and therefore de-identification) purposes. This corpus includes more than 100 speakers with various accents[3] who recorded around 400 utterances each. These utterances include the Rainbow Passage [89], an elicitation paragraph, and sentences selected from a newspaper, the latter being different for each speaker. Since the techniques used in these experiments require a parallel corpus for training VC functions, the elicitation paragraph plus the Rainbow Passage were used for training, whereas the newspaper sentences were used for testing. It must be noted that not all the training sentences are available for all the speakers, so the amount of training utterances for each speaker differs slightly.

The second dataset used in these experiments is that of the Voice Conversion Challenge 2016 (VCC 2016) [90]. This corpus is based on the Data and Production Speech dataset [91], a freely available corpus recorded by professional US English speakers in a recording studio. Specifically, the "clean" version of the dataset was used in VCC 2016, and it includes around 13 min of speech, which were split into train and test sets, from each of the ten speakers that were selected for this corpus. The utterances include sentences from public domain books (novels such as *Alice's Adventures in Wonderland*, *Twenty Thousand Leagues Under the Seas*, and

Table 5.2   Experimental framework

	VCTK	VCC 2016
# of speakers	109	10
Average # of training utterances	23	162
Average # of test utterances	383	54
Average duration training utterances (per speaker)	2 min 30 s	9 min 41 s
Average duration test utterances (per speaker)	21 min 44 s	2 min 57 s

---

[3]The accents featured in VCTK Corpus are American, Australian, Canadian, English, Indian, Irish, New Zealand, Northern Irish, Scottish, South African, and Welsh.

Treasure Island, among others). All the speakers recorded the same sentences, so there is a parallel corpus for every pair of speakers.

Some statistics of VCTK and VCC 2016 corpora, as used in these experiments, are summarized in Table 5.2.

5.5.1 Evaluation metrics

As mentioned in Section 5.2, there are different techniques and metrics to evaluate the performance of speaker de-identification strategies. In these experiments, objective metrics were chosen to assess both the de-identification accuracy and the intelligibility of the resulting speech.

De-identification was measured by means of the equal error rate (EER), accompanied by detection error trade-off (DET) curves. The EER represents the error at the operating point where the false reject probability is equal to the false accept probability, whereas the DET curve is a plot of the false alarm and miss probabilities at different decision thresholds. These metrics were obtained employing an automatic speaker verification system. For this purpose, the experimental procedure described by Lopez-Otero *et al.* [43] was followed. First, the hyper-parameters of the speaker verification system were tuned using the original speech, i.e., speaker verification experiments were run on the two datasets varying the hyper-parameters to find those that led to the lowest EER. Then, experiments were run using the de-identified data: in this case, the enrollment utterances were original speech, whereas the test utterances comprised de-identified speech, and the goal is to achieve an EER as high as possible. The experimental procedure defined by Lopez-Otero *et al.* [43] was followed in these experiments. First, all the test utterances of each speaker were concatenated into a single one and subsequently divided in utterances of 15 s of duration. The first utterance was used for enrollment, whereas the others were used for testing.

The speaker verification system used in these experiments employs i-vectors for speech representation and PLDA scoring [82]. This system, implemented using the Kaldi toolkit [92], requires a UBM, a total variability matrix, and a PLDA transformation matrix. The data used to train the system was extracted from the DS1 partition of the Biosecure database, which is described in Section 5.4.2. The UBM is a GMM with a diagonal covariance matrix, the total variability matrix was trained following Dehak *et al.* [81], and the features used are 19 MFCCs augmented with their delta and acceleration coefficients. The number of mixtures of the UBM and the dimension of the i-vectors were tuned as described above.

The intelligibility of the de-identified speech was measured by computing the WER achieved after applying a state-of-the-art large vocabulary ASR engine. The ASR is based on the Kaldi toolkit and uses hybrid time-delay neural network (TDNN)–Hidden Markov Model acoustic models with three-gram and four-gram language models. Specifically, the DNN acoustic model is a factorized TDNN with skip connections whose architecture consists of 16 time-delay hidden layers of dimension 1 536 factorized with a linear bottleneck dimension of 160. The network input features are 40-dimensional MFCC appended with a 100-dimensional i-vector for speaker adaption. The data used for training the models used on the ASR engine

Table 5.3   EER and WER achieved on VCTK corpus with different de-identification techniques in speaker-dependent and speaker-independent settings. Results on original speech are shown for comparison

Experiment	Speaker-dependent (%)		Speaker-independent (%)	
	EER	**WER**	**EER**	**WER**
Original	4.00	9.45	4.00	9.45
FW + AS piecewise ($|\alpha|=$ $\pi/24$)	42.59	17.46	42.59	17.46
FW + AS pretrain	57.01	13.14	53.86	13.55
DNN	67.33	42.59	61.90	55.14
GAN	59.08	34.25	52.03	36.64

Table 5.4   EER and WER achieved on VCC2016 corpus with different de-identification techniques in speaker-dependent and speaker-independent settings. Results on original speech are shown for comparison

Experiment	Speaker-dependent (%)		Speaker-independent (%)	
	EER	**WER**	**EER**	**WER**
Original	0.98	6.05	0.98	6.05
FW + AS piecewise ($|\alpha|=$ $\pi/24$)	55.88	8.70	55.88	8.70
FW + AS pretrain	59.8	6.87	62.75	6.93
DNN	69.61	15.19	69.61	19.35
GAN	69.61	12.92	60.78	16.15

was the LibriSpeech corpus of audiobooks [93]. In particular, the acoustic models were trained using the 1 000 hours LibriSpeech training set, and the models freely available online[4] were used as language models. The ASR engine uses a two-pass decoding strategy; in the first decoding, it uses a small three-gram language model to generate a lattice, which in the second pass is rescored with the large four-gram language model in order to get the output transcription.

---

[4]http://www.openslr.org/11/

5.6 Results and analysis

Tables 5.3 and 5.4 present the results achieved in VCTK and VCC 2016 corpora, respectively, in terms of EER and WER. These tables summarize the experiments in speaker-dependent and speaker-independent settings. It must be noted that the FW + AS piecewise transformation is intrinsically speaker-independent, so the same results are displayed for both sets of experiments. The tables also show the performance achieved for original speech where, as in the FW + AS piecewise, the result is the same for speaker-independent and speaker-dependent settings.

First, the EER of the original speech was computed to assess the validity of the speaker verification system used in these experiments. As explained above, the hyper-parameters of the system were tuned on these data, leading to UBM number of mixtures and dimension of i-vectors equal to 1 024 and 200 for VCTK corpus and equal to 512 and 400 for VCC 2016. The obtained EERs using original speech were 4% and 1% in VCTK and VCC 2016 corpora, respectively, which is an acceptable result in the experimental settings defined for these experiments (see [94] for state-of-the-art speaker verification results). Also, these EER values serve as a reference to compare with the EER achieved on de-identified speech.

The FW + AS piecewise system exhibited the expected performance given the speaker de-identification experiments in previous works [11, 43, 44]. The results show that the lower the value of $|\alpha|$ the greater the EER since smaller values of $|\alpha|$ lead to coarser conversion functions and, therefore, the perceived de-identified speaker is less similar to the original speaker.

There is also previous work describing speaker de-identification results using the FW + AS pretrain strategy [16, 43], which are coherent with those displayed in Tables 5.3 and 5.4. It can be observed that this technique achieves an EER similar to that observed for FW + AS piecewise with $|\alpha| = \pi/24$. Nevertheless, the WER is 4% lower than that achieved with the piecewise strategy. The FW + AS pretrain system can be used in speaker-dependent and speaker-independent settings, which allows the comparison of the different results achieved in these conditions. It can be shown that there is no agreement on which setting leads to better de-identification since the speaker-dependent condition exhibited a higher EER than the speaker-independent one in the VCTK corpus, but the contrary happened in VCC 2016. This might be caused by the small number of source and target speakers in the VCC 2016 corpus, which may lead to source speakers that are not very similar to the input speakers, and to target speakers that are not very dissimilar to the source speakers. In terms of intelligibility, speaker-dependent and speaker-independent techniques showed WERs that are not significantly different.

After commenting on the results achieved with the FW + AS techniques, the performance of the systems based on deep learning techniques must be discussed. Tables 5.3 and 5.4 show that, in general, the best de-identification results are achieved with the DNN system. The speaker de-identification performance observed in the speaker-dependent setting was greater than that of the speaker-independent setting for the VCTK corpus (as with the FW + AS pretrain system), whereas they were

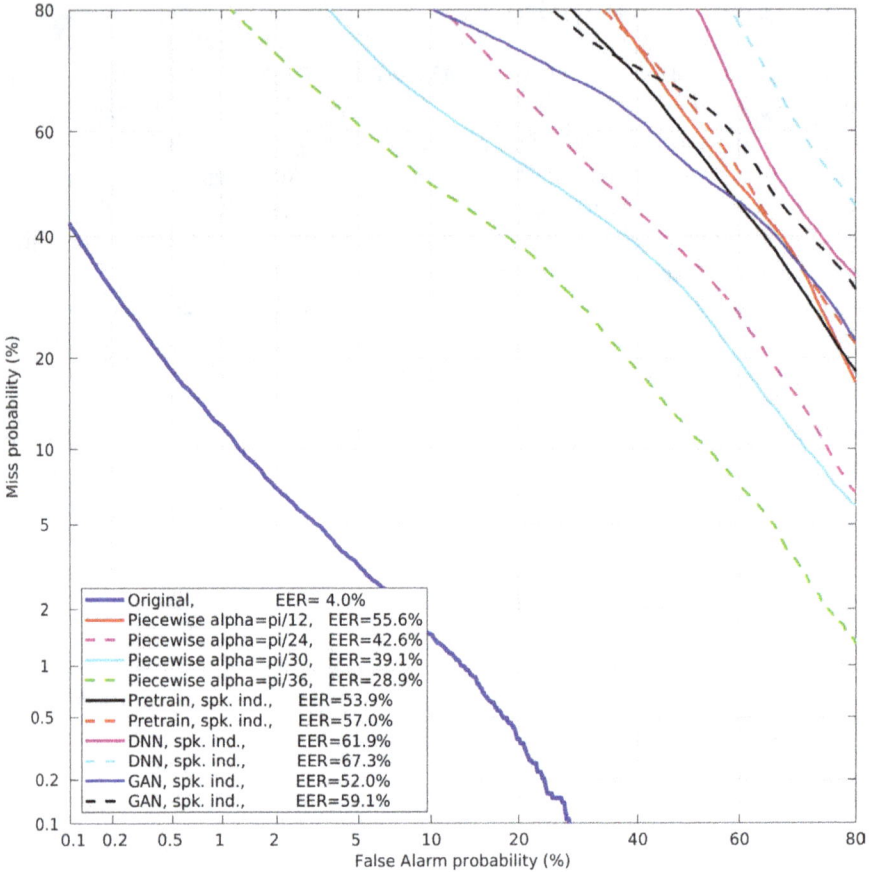

Figure 5.3   DET curves of the different systems on the VCTK corpus

equal for the VCC 2016 corpus. However, the WER was significantly higher than the values exhibited for the FW + AS techniques, which suggests that the resulting de-identified speech is not so intelligible according to the ASR performance.

Last, the GAN system showed competitive results in terms of EER in both datasets and, in this case, the results observed for speaker-independent and speaker-dependent settings were equivalent, having a higher EER when performing speaker-dependent de-identification. Even though the WERs obtained with this strategy were not as high as those obtained with the DNN approach, GAN systems are far from FW + AS strategies in terms of intelligibility. This poor intelligibility performance highlights that, to achieve a good quality of the converted speech, deep-learning-based techniques still suffer from dependence on a large amount of training data.

Figures 5.3 and 5.4 show the DET curves of the different systems on VCTK and VCC 2016 corpora, respectively. These DET curves show that, even though the

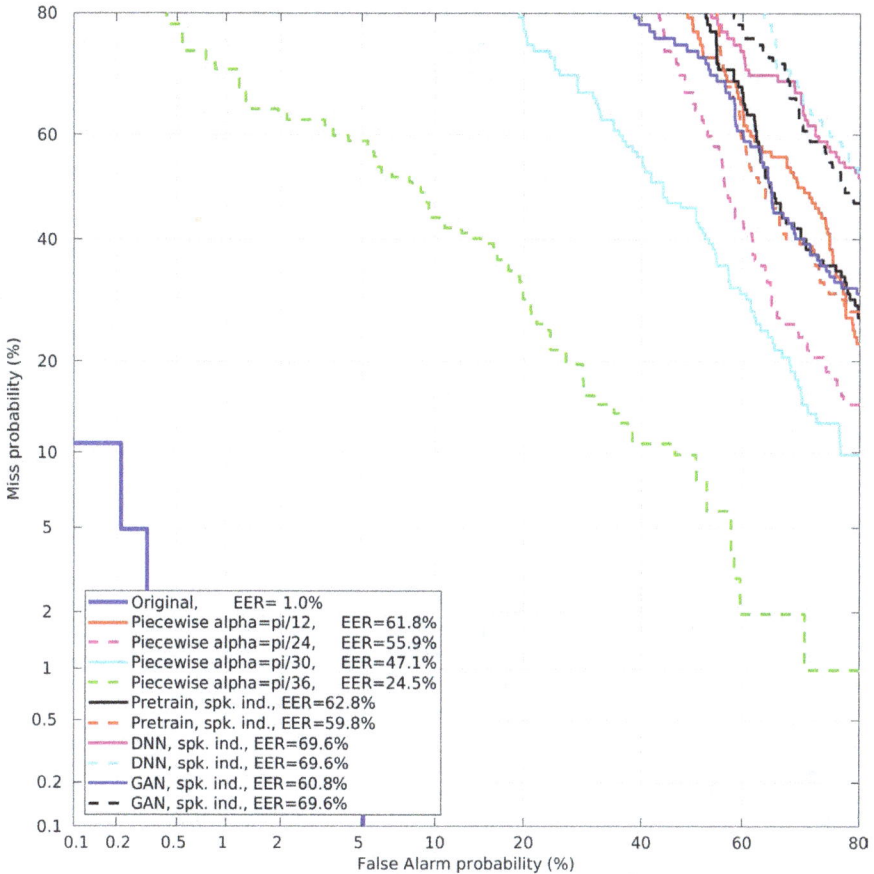

Figure 5.4   DET curves of the different systems on the VCC 2016 corpus

different systems differ in performance, their EERs are far from those achieved with original speech, which suggests that speaker de-identification is successful with all the strategies.

5.7   Conclusion

In today's world where society is increasingly concerned about the use of personal data, having high-performance speaker de-identification technology is of great interest to the public in general and companies in particular. Therefore, this chapter addresses the relevant topic of how to eliminate information about the speaker while maintaining the speech-like waveform.

As stated in the preceding Sections 5.1 and 5.3, speaker de-identification technology is based primarily on VC technology. This increases its scope, making it not such a narrow research area and giving space to constant innovation.

The experimental results described in Section 5.4 compare classical techniques based on FW + AS with modern approaches based on deep learning. Objective performance measures showed good de-identification performance, since the EER obtained when dealing with de-identified speech dramatically increased compared with that achieved with original speech. In addition, the obtained WER suggests that the intelligibility of the resulting de-identified speech is good but still has room for improvement, especially in the case of deep learning techniques when limited amount of training data is available.

In conclusion, FW + AS techniques guarantee very small quality degradation, and also the possibility of recovering the original speech (re-identification) by inverting the transformation. Moreover, the computational cost of the proposed approach is small, which allows the system to process speech in real time. Deep learning techniques are promising in terms of universality and naturalness but, at this moment, there are no studies reporting re-identification performance using DNN-based de-identification approaches.

Acknowledgments

This work was partially supported by the Spanish Ministerio de Economia y Competitividad through the project Speech&Sign RTI2018-101372-B-100, and also by Xunta de Galicia (AtlanTTic and ED431B 2018/60 grants) and the European Regional Development Fund (ERDF).

References

[1]   Garfinkel S.L. 'De-identification of personally identifiable information. National Institute of standards and technology (NIST)'. *US. Department of Commerce.* 2015.

[2]   Meystre S.M., Ferrández O., Friedlin F.J., South B.R., Shen S., Samore M.H. 'Text de-identification for privacy protection: a study of its impact on clinical text information content'. *Journal of Biomedical Informatics.* 2014;50(1):142–50.

[3]   Obeid J.S., Heider P.M., Weeda E.R., *et al.* 'Impact of de-identification on clinical text classification using traditional and deep learning classifiers'. *Studies in Health Technology and Informatics.* 2019;264:283–7.

[4]   Friedrich M., Köhn A., Wiedemann G. 'Adversarial learning of privacy-preserving text representations for de-identification of medical records'. *Proceedings of the 57th Annual Meeting of the Association for Computational Linguistics*; 2019. pp. 5829–39.

[5]   Newton E.M., Sweeney L., Malin B. 'Preserving privacy by de-identifying face images'. *IEEE Transactions on Knowledge and Data Engineering.* 2005;17(2):232–43.

[6]   Gross R., Sweeney L., Cohn J. 'Face de-identification'. *Protecting Privacy in Video Surveillance.* Springer Publishing Company, Incorporated; 2009. pp. 129–46.

[7] Agrawal P., Narayanan P.J. 'Person de-identification in videos'. *IEEE Transactions on Circuits and Systems for Video Technology.* 2011;21(3):299–310.

[8] Gafni O., Wolf L., Taigman Y. 'Live face de-identification in video'. *The IEEE International Conference on Computer Vision (ICCV)*; 2019. pp. 9378–87.

[9] Jin Q., Toth A.R., Schultz T., *et al.* 'Speaker de-identification via voice transformation'. IEEE Workshop on Automatic Speech Recognition and Understanding; 2009. pp. 529–33.

[10] Pobar M., Ipsic I. 'Online speaker de-identification using voice transformation'. 37th International Convention on Information and Communication Technology, Electronics and Microelectronics (MIPRO); 2014. pp. 1264–7.

[11] Magariños C., Lopez-Otero P., Docio-Fernandez L., *et al.* 'Piecewise linear definition of transformationfunctions for speaker de-identification'. Proceedings of First International Workshop on Sensing, Processing and Learning for Intelligent Machines (SPLINE); 2016. pp. 1–5.

[12] Orozco-Arroyave J.R., Hönig F., Arias-Londoño J.D., *et al.* 'Automatic detection of Parkinson's disease in running speech spoken in three different languages'. *The Journal of the Acoustical Society of America.* 2016;139(1):481–500.

[13] König A., Satt A., Sorin A., *et al.* 'Automatic speech analysis for the assessment of patients with predementia and Alzheimer's disease'. *Alzheimer's & Dementia: Diagnosis, Assessment & Disease Monitoring.* 2015;1(1):112–24.

[14] Karam Z.N., Provost E.M., Singh S., *et al.* 'Ecologically valid long-term mood monitoring of individuals with bipolar disorder using speech'. ICASSP; 2014. pp. 4858–62.

[15] Cummins N., Epps J., Breakspear M., *et al.* 'An investigation of depressed speech detection: features and normalization'. Interspeech; 2011. pp. 2997–3000.

[16] Magariños C., Lopez-Otero P., Docio-Fernandez L., Rodriguez-Banga E., Erro D., García-Mateo C. 'Reversible SPEAKER de-identification using pre-trained transformation functions'. *Computer Speech and Language.* 2017;46(12):36–52.

[17] Ribaric S., Ariyaeeinia A., Pavesic N. 'De-identification for privacy protection in multimedia content: a survey'. *Signal Processing: Image Communication.* 2016;47:131–51.

[18] Dall R., Yamagishi J., King S. 'Rating naturalness in speech synthesis: the effect of style and expectation'. Proc. Speech Prosody; 2014.

[19] Desai S., Raghavendra E.V., Yegnanarayana B., *et al.* 'Voice conversion using artificial neural networks'. Proceedings of ICASSP; 2009. pp. 3893–6.

[20] Wester M., Wu Z., Yamagishi J. 'Analysis of the voice conversion challenge 2016 evaluation results'. Interspeech; 2016.

[21] Lorenzo-Trueba J., Yamagishi J., Toda T., *et al.* 'The voice conversion challenge 2018: promoting development of parallel and nonparallel methods'. Odyssey 2018. The Speaker and Language Recognition Workshop; 2018. pp. 195–202.

[22] Niwa J., Yoshimura T., Hashimoto K., *et al.* 'Statistical voice conversion based on wavenet'. ICASSP; 2018. pp. 5289–93.

[23] Chen K., Chen B., Lai J., *et al.* 'High-quality voice conversion using spectrogram-based Wavenet vocoder'. Interspeech; 2018. pp. 1993–7.

[24] Bahmaninezhad F., Zhang C., Hansen J.H.L. 'Convolutional neural network based speaker de-identification'. Odyssey 2018. The Speaker and Language Recognition Workshop; 2018. pp. 255–60.

[25] Toda T., Black A.W., Tokuda K. 'Voice conversion based on maximum-likelihood estimation of spectral parameter trajectory'. *IEEE Transactions on Audio, Speech and Language Processing.* 2007;15(8):2222–35.

[26] Jin Z., Finkelstein A., andLu D.J.S., *et al.* 'CUTE: a concatenative method for voice conversion using examplar-based unit selection'. ICASSP; 2016. pp. 5660–4.

[27] Tian X., Wang J., Xu H., *et al.* 'Average modeling approach to voice conversion with non-parallel data'. Odyssey 2018. The Speaker and Language Recognition Workshop; 2018. pp. 227–32.

[28] Sun L., Li K., Wang H., *et al.* 'Phonetic posteriorgrams for many-to-one voice conversion without parallel data training'. IEEE International Conference on Multimedia and Expo; 2016.

[29] Lee K.-S. 'Restricted Boltzmann machine-based voice conversion for non-parallel corpus'. *IEEE Signal Processing Letters.* 2017;24(8):1103–7.

[30] Veaux C., Yamagishi J., King S. 'Using HMM-based speech synthesis to reconstruct the voice of individuals with degenerative speech disorders'. Interspeech; 2012. pp. 967–70.

[31] Nakamura K., Toda T., Saruwatari H. 'Speaking-aid systems using GMM-based voice conversion for electrolaryngeal speech'. *Speech Communication.* 54; 2012. pp. 134–46.

[32] Toda T., Nakagiri M., Shikano K. 'Statistical voice conversion techniques for body-conducted unvoiced speech enhancement'. *IEEE Transactions on Audio, Speech, and Language Processing.* 2012;20(9):2505–17.

[33] Taguchi F., Kaburagi T. 'Articulatory-to-speech conversion using bi-directional long short-term memory'. Interspeech; 2018. pp. 2499–503.

[34] Wolters M.K., Isaac K.B., Renals S. 'Evaluating speech synthesis intelligibility using Amazon Mechanical Turk'. Proc. 7th Speech Synthesis Workshop (SSW7); 2010. pp. 136–41.

[35] Ullmann R., Rasipuram R., Magimai-Doss M., *et al.* 'Objective intelligibility assessment of text-to-speech systems through utterance verification'. Interspeech; 2016. pp. 3501–5.

[36] Shah N., Shah N.J., Patil H.A. 'Effectiveness of generative adversarial network for non-audible murmur-to-whisper speech conversion'. Interspeech; 2018. pp. 3157–61.

[37] Benoît C., Grice M., Hazan V. 'The SUS test: a method for the assessment of text-to-speech synthesis intelligibility using semantically unpredictable sentences'. *Speech Communication.* 1996;18(4):381–92.

[38] Justin T., Mihelič F., Dobrišek S. 'Intelligibility assessment of the de-identified speech obtained using phoneme recognition and speech synthesis systems'. *Text, Speech and Dialogue. vol. 8655 of Lecture Notes in Computer Science.* Springer; 2014. pp. 529–36.

[39] Toth A.R., Black A.W. 'Using articulatory position data in voice transformation'. Proc. 6th ISCA Workshop on Speech Synthesis (SSW6); 2007. pp. 182–7.

[40] Abou-Zleikha M., Tan Z.H., Christensen M.G., *et al.* 'A discriminative approach for speaker selection in speaker de-identification systems'. Proceedings of 23rd European signal processing conference (EUSIPCO); 2015. pp. 2147–51.

[41] Jin Q., Schultz T., Black A. 'Voice convergin: speaker de-identification by voice transformation'. ICASSP; 2009. pp. 3909–12.

[42] Justin T., truc V., Dobrišek S., *et al.* 'Speaker de-identification using diphone recognition and speech synthesis'. Conference and Workshops on Automatic Face Gesture Recognition; 2015. pp. 1–7.

[43] Lopez-Otero P., Magariños C., Docio-Fernandez L., *et al.* 'Influence of speaker de-identification in depression detection'. IET Signal Processing; 2017.

[44] Lopez-Otero P., Docio-Fernandez L., Abad A., *et al.* 'Depression detection using automatic transcriptions of de-identified speech'. Interspeech; 2017. pp. 3157–61.

[45] Wolters M.K., Isaac K.B. 'Crowdsourcing speech intelligibility judgments'. *The Journal of the Acoustical Society of America.* 2017;141(5):3911.

[46] Fukumoto M. 'SilentVoice: unnoticeable voice input by ingressive speech'. Proc. UIST; 2018. pp. 2499–503.

[47] Serrano L., Tavarez D., Sarasola X., *et al.* 'LSTM based voice conversion for laryngectomees'. Iberspeech; 2018. pp. 122–6.

[48] Nurminen J., Silén H., Popa V., *et al.* 'Voice Conversion' in Ramakrishnan S. (ed.). *Speech Enhancement, Modeling and Recognition – Algorithms and Applications.* Rijeka: IntechOpen; 2012.

[49] Wu H., Wang Y., Huang J. 'Blind detection of electronic disguised voice'. ICASSP; 2013. pp. 3013–17.

[50] Perrot P., Aversano G., Chollet G. 'Voice disguise and automatic detection: review and perspectives'. *Progress in Nonlinear Speech Processing. vol. 4391 of Lecture Notes in Computer Science.* Springer; 2007. pp. 101–17.

[51] Mohammadi S.H., Kain A. 'An overview of voice conversion systems'. *Speech Communication.* 2017;88(3):65–82.

[52] Mouchtaris A., Van der Spiegel J., Mueller P. 'Non-parallel training for voice conversion by maximum likelihood constrained adaptation'. ICASSP; 2004. pp. I–1.

[53] Sündermann D., Bonafonte A., Höge H., *et al.* 'A first step towards text-independent voice conversion'. Interspeech; 2004.

[54] Sündermann D. *Text-independent Voice Conversion.* Universitätsbibliothek der Universität der Bundeswehr München; 2008.

[55] Sündermann D., Ney H. 'VTLN-based voice conversion'. Proceedings of ISSPIT; 2003. pp. 556–9.

[56] Türk O. *Cross-Lingual Voice Conversion*. Bogazici University; 2007.

[57] Abe M., Nakamura S., Shikano K., *et al.* 'Voice conversion through vector quantization'. Proceedings of the IEEE International Conference on Acoustics, Speech, Signal Processing (ICASSP); 1988. pp. 565–8.

[58] Türk O., Arslan L.M. 'Robust processing techniques for voice conversion'. *Computer Speech & Language.* 2006;20(4):441–67.

[59] Stylianou Y., Cappé O., Moulines E. 'Continuous probabilistic transform for voice conversion'. *IEEE Transactions on Speech and Audio Processing.* 1998;6(2):131–42.

[60] Erro D., Moreno A., Bonafonte A. 'Voice conversion based on weighted frequency warping'. *IEEE Transactions on Audio, Speech, and Language Processing.* 2010;18(5):922–31.

[61] Zorila T., Erro D., Hernaez I. 'Improving the quality of standard GMM-based voice conversion systems by considering physically motivated linear transformations'. *Advances in Speech and Language Technologies for Iberian Languages.* Springer; 2012. pp. 30–9.

[62] Desai S., Black A.W., Yegnanarayana B., Prahallad K., *et al.* 'Spectral mapping using artificial neural networks for voice conversion'. *IEEE Transactions on Audio, Speech, and Language Processing.* 2010;18(5):954–64.

[63] Hsu C.C., Hwang H.T., YC W., *et al.* 'Voice conversion from non-parallel corpora using variational auto-encoder'. 2016 Asia-Pacific Signal and Information Processing Association Annual Summit and Conference (APSIPA); 2016.

[64] Shikano K., Nakamura S., Abe M. 'Speaker adaptation and voice conversion by codebook mapping'. IEEE International Symposium on Circuits and Systems; 1991. pp. 594–7.

[65] Arslan L.M. 'Speaker transformation algorithm using segmental Codebooks (STASC'. *Speech Communication.* 1999;28(3):211–26.

[66] Kain A., Macon M.W. 'Spectral voice conversion for text-to-speech synthesis'. ICASSP; 1998.

[67] Takamichi S., Toda T., Black A.W., Neubig G., Sakti S., Nakamura S. 'Postfilters to modify the modulation spectrum for statistical parametric speech synthesis'. *IEEE/ACM Transactions on Audio, Speech, and Language Processing.* 2016;24(4):755–67.

[68] Pitz M., Ney H. 'Vocal tract normalization equals linear transformation in cepstral space'. *IEEE Transactions on Speech and Audio Processing.* 2005;13(5):930–44.

[69] Erro D., Navas E., Hernaez I. 'Parametric voice conversion based on bilinear frequency Warping plus amplitude scaling'. *IEEE Transactions on Audio, Speech, and Language Processing.* 2013;21(3):556–66.

[70] Godoy E., Rosec O., Chonavel T. 'Voice conversion using dynamic frequency warping with amplitude scaling, for parallel or nonparallel

corpora'. *IEEE Transactions on Audio, Speech, and Language Processing.* 2012;20(4):1313–23.

[71] Tamura M., Morita M., Kagoshima T., *et al.* 'One sentence voice adaptation using GMM-based frequency-warping and shift with a sub-band basis spectrum model'. 2011 IEEE International Conference on Acoustics, Speech and Signal Processing (ICASSP); 2011. pp. 5124–7.

[72] Narendranath M., Murthy H.A., Rajendran S., Yegnanarayana B., *et al.* 'Transformation of formants for voice conversion using artificial neural networks'. *Speech Communication.* 1995;16(2):207–16.

[73] Nakashika T., Takashima R., Takiguchi T., *et al.* 'Voice conversion in high-order eigen space using deep belief nets'. INTERSPEECH 2013, 14th Annual Conference of the International Speech Communication Association; 2013. pp. 369–72.

[74] Chen L.H., Ling Z.H., Song Y., *et al.* 'Joint spectral distribution modeling using restricted Boltzmann machines for voice conversion'. Interspeech; 2013. pp. 3052–6.

[75] Saito Y., Ijima Y., Nishida K., *et al.* 'Non-parallel voice conversion using variational autoencoders conditioned by phonetic posteriorgrams and d-vectors'. ICASSP; 2018. pp. 5274–8.

[76] Hsu W.N., Zhang Y., Glass J. 'Learning latent representations for speech generation and transformation'. Interspeech; 2017. pp. 1273–7.

[77] Erro D., Alonso A., Serrano L., Navas E., Hernaez I., *et al.* 'Interpretable parametric voice conversion functions based on Gaussian mixture models and constrained transformations'. *Computer Speech & Language.* 2015;30(1):3–15.

[78] Erro D., Sainz I., Navas E., *et al.* 'Improved HNM-based vocoder for statistical synthesizers'. Interspeech; 2011. pp. 1809–12.

[79] Reynolds D.A., Quatieri T.F., Dunn R.B. 'Speaker verification using adapted Gaussian mixture models'. *Digital Signal Processing.* 2000;10(1–3):19–41.

[80] Moreno A., Poch D., Bonafonte A., *et al.* 'Albayzin speech database: design of the phonetic corpus'. EUROSPEECH. vol. 1; 1993. pp. 175–8.

[81] Dehak N., Kenny P.J., Dehak R., *et al.* 'Front end factor analysis for speaker verification'. IEEE Transactions on Audio, Speech and Language Processing; 2010.

[82] Garcia-Romero D., Espy-Wilson C.Y. 'Analysis of i-vector length normalization in speaker recognition systems'. Proceedings of Interspeech; 2011. pp. 249–52.

[83] Ortega-Garcia J., Fierrez J., Alonso-Fernandez F., *et al.* 'The Multiscenario Multienvironment BioSecure multimodal database (BMDB)'. *IEEE Transactions on Pattern Analysis and Machine Intelligence.* 2009;32(4):1097–111.

[84] Wu Z., Watts O., King O. 'Merlin: an open source neural network speech synthesis system'. 9th ISCA Speech Synthesis Workshop; 2016. pp. 202–7.

[85] Morise M., Yokomori F., Ozawa K. 'World: a vocoder-based high-quality speech synthesis system for real-time applications'. *IEICE Transactions on Information and Systems.* 2016;E99.D(7):1877–84.

[86]   SPTK Working Group. The SPTK toolkit. Version 3.9. Available from http://
       sp-tk.sourceforge.net/.

[87]   Saito Y., Takamichi S., Saruwatari H. 'Statistical parametric speech synthesis
       incorporating generative adversarial networks'. *IEEE/ACM Transactions on
       Audio, Speech, and Language Processing*. 2018;26(1):84–96.

[88]   Veaux C., Yamagishi J., MacDonald K. *CSTR VCTK corpus: English multi-
       speaker corpus for CSTR voice cloning toolkit*. University of Edinburgh. The
       Centre for Speech Technology Research (CSTR); 2016.

[89]   Fairbanks G. *Voice and Articulation Drillbook*. 2nd ed. New York: Harper &
       Row; 1960.

[90]   Toda T., Chen L.H., Saito D., *et al.* 'The voice conversion challenge 2016'.
       Interspeech. vol. 1; 2016. pp. 1632–6.

[91]   Mysore G.J. 'Can we automatically transform speech recorded on common
       consumer devices in real-world environments into professional production
       quality speech? – a dataset, insights, and challenges'. *IEEE Signal Processing
       Letters*. 2015;22(8):1006–10.

[92]   Povey D., Ghoshal A., Boulianne G., *et al.* 'The Kaldi speech recogni-
       tion toolkit'. IEEE Workshop on Automatic Speech Recognition and
       Understanding (ASRU); 2011.

[93]   Panayotov V., Chen G., Povey D., *et al.* 'Librispeech: An ASR corpus based
       on public domain audio books'. 2015 IEEE International Conference on
       Acoustics, Speech and Signal Processing (ICASSP); 2015. pp. 5206–10.

[94]   Sadjadi S.O., Kheyrkhah T., Tong A., *et al.* 'The 2016 NIST Speaker
       Recognition Evaluation'. Interspeech; 2017. pp. 1353–7.

Chapter 6

Performance evaluation of voice biometrics solutions

Jean-François Bonastre[1] and Anthony Larcher[2]

This chapter is dedicated to performance evaluation of voice biometrics solutions. The specific aspects of voice biometrics, compared to other speech-based technologies, are presented, as well as their consequences in terms of performance evaluation. The main existing evaluation protocols and metrics are then presented and discussed, with a focus on the speaker verification part of a voice biometrics systems. Other aspects such as calibration, diarization and forensic or privacy are also introduced. Finally, some limits of performance evaluation and some guidelines are proposed.

6.1 Introduction

There is no doubt about the fact that the technology should be seriously evaluated before it is used in the real world. If this statement is true for all voice-based applications, it takes a greater importance with voice biometrics area. Three main reasons could explain this fact:

- First, when an error in speech recognition will generally simply lead to a second attempt by the user, a false acceptance error in voice biometrics (a fraudster accepted as a registered customer, for example) may have severe consequences.
- Second, voice biometrics is also linked to the fields of security and forensics, where the cost of an error can be expressed not in money but in terms of human life.
- Third, speech biometrics inherently deals with the differences between human voices. Unfortunately, it is sometimes linked to differences between humans,

[1]Laboratoire Informatique d'Avignon (LIA), University of Avignon, Avignon, France
[2]Laboratoire d'Informatique de l'Université du Mans (LIUM), University of Le Mans, Le Mans, France

when the mother tongue or regional accent is mentioned, for example. The border between modeling differences and discrimination can be narrow, so that speech biometrics is a more sensitive area than other speech technologies, as potential biases in learning datasets or algorithms could lead to discrimination issues.

Evaluating speech biometric solutions is also intrinsically more difficult than for other speech-based technologies, such as speech recognition systems, as speech biometrics is particularly sensitive to variability and nuisance factors, two elements that affect all speech-based applications. Variability and nuisance come from background noises, microphones, transmission channels, spoken language, regional accents, speaking style, pragmatic situation, linguistic content of the voice recording, emotional state of the speaker, the speaker her/himself, etc. However, the influence of variability and nuisance factors when evaluating technological solutions for voice biometrics is different for three main reasons [1]:

- First, as already said in the introduction, voice biometrics (more precisely, its mother scientific field, speaker recognition) plays with the differences and similarities between individuals, with the idea that the differences in audio signal between individuals are larger than the differences for a given individual. This means that the assessment of speech biometrics requires careful consideration of intra-speaker variability, with a lot of data per speaker, whereas, on the other hand, assessment of speech recognition only requires covering the space of variability using all data from all speakers.
- Second, when for an application such as speech transcription, increasing the level of variability and nuisance (switching from mono- to multi-speakers or decreasing the signal-to-noise ratio for certain speech extracts, for example) always leads to decreased performance; this is not always the case for voice biometrics. In voice biometrics, adding a new variability factor to an experimental dataset in order to increase the difficulty can, on the contrary, help an automatic system to obtain better general performances for this experiment. For example, adding a new microphone to a test set can increase performance if few speakers use that microphone. Idem if few new speakers are recorded and bring to the database new regional accents or mother languages.
- Third, biometric voice applications always rely on a score normalization step applied on the outputs of the core speaker recognition engine. Two main families of score normalization are used. The first family computes score statistics (e.g., mean and standard deviation) at the level of individual test utterances (T-norm) or speakers (Z-norm) in order to output a standardized detection score. The second family of approaches is usually called "calibration" and refers to global, order-preserving transforms that aim to convert raw scores into calibrated log-likelihood ratios. In the context of performance evaluation, the calibration parameters (and, in some situations, the score normalisation parameters) are often calculated a posteriori to simplify the process. This solution must be

taken with caution because it can mask certain weaknesses of the basic recognition engine and also accepts biases and intrinsic problems (as shown in [2]).

Usually, the cooperation of users of a voice biometric system is taken for granted. This includes the cooperation of a fraudster, who wishes to access it and meets the system requirements. The generalization of a performance evaluation carried out with this paradigm of cooperation to a case where the users do not cooperate should not be done. This is particularly important for forensic applications, where it is obvious that the speaker's cooperation cannot be taken for granted: generally, forensic voice comparison analysis are requested by the prosecution and not the defense, so a suspect has little interest in cooperating with the system. More generally, the forensic aspects must be taken with caution [3, 4], and the performance observed in a more traditional commercial configuration must never be transferred as is to this field.

Since several years, the possibility to spoof a voice biometrics system using artificial transformation of the voice has been developing [5, 6]. Thus, the question of the robustness of voice biometrics solutions to technological attacks such as conversion or speech synthesis becomes very important and, obviously, this robustness must be evaluated. This aspect is however not directly linked to voice biometrics, and specific evaluation protocols had to be developed for this purpose [7].

This chapter is mainly dedicated to how the voice biometrics core engine, the speaker verification system, is evaluated.

6.2 Evaluating methods or technology

Creating protocols, defining metrics and collecting appropriate data to evaluate speaker recognition is a tedious, complex and expensive task that has been mobilizing the entire scientific community for decades through benchmarking evaluations.

Those evaluations should be considered with respect to different factors as their implementation and rules depend on the goals of the sponsoring organizations, but they widely benefit to the scientific community by providing public material and by emulating the discussions and criticisms of the evaluation frameworks. The first benefit of those evaluations consists of the large quantities of data produced and released over the years, increasing the number of speech segments (a speech segment here refers to a voice recording or part of a voice recording used to compute a voice authentication or to enroll a speaker) from a few hundreds to several thousands, as well as the number of speakers. Taken together, the number of segments and speakers increases the number of trials from a few hundreds to tens of millions and allows for greater significance of performance measures. The diversity of conditions involved in the databases has also increased over the decades, helping to generalize the results of the performance evaluation. The second impact of those evaluation is to develop protocols, improve metrics and push the community to higher evaluation standards.

6.2.1   *Existing benchmarking evaluations*

Speaker recognition has a long history of international benchmarking evaluations starting in the 1990s. Along the years, the quality and the complexity of the tasks have greatly improved. If the NIST[1] has been the major actor in this domain, with 17 evaluations organized over 23 years [8], other organizations have conducted benchmarking evaluation in order to attract the attention of the research community on different aspects or conditions of the tasks as the focus of evaluation remains a decision of the sponsors.

From 1996 to 2006, the National Institute of Standards and Technology (NIST) has led yearly evaluations, before reducing the pace to biyearly as the complexity of the evaluation increased for both organizers and participants. NIST speaker recognition evaluations (SREs) mainly focus on speaker verification on telephone, but conditions and tasks evolved across time.

Conditions of the evaluations have evolved over the years to focus on different known limitations of the technology, essentially due to the sources of variability like the transmission channel and noise, the duration of recordings, the language mismatch or the aging of the speakers.

The transmission channel has been addressed over the years by recording the speech segments used for NIST SRE over different types of telephones (landlines, cellular, VoIP) and microphones (close, distant, interview) [9]. However, evaluation data mostly consists of telephone speech recorded in 8 kHz[2] through the North American communication network. Starting from 2016, data from other countries has been introduced massively together with Voice over Internet Protocol (VoIP) channels. NIST SRE 2016 has also seen the introduction of recordings including real background noises as a specific part of the challenge.

Duration of the recording has always been a challenge for automatic systems, and NIST evaluations have been conducted for years with a focus on different fixed speech durations ranging from 10 seconds to several minutes before moving to more variability in 2016. The impact of language mismatch is another factor that has been evaluated in this series of challenges from the early 2000s [8]. The effects of bilingual speakers (SRE 2004/2008, with speakers alternately speaking English and one other language between Arabic, Mandarin, Russian or Spanish) and an unknown language (SRE 2016) have both been studied. This series of evaluations organized along the years has also produced data and baselines for the study of speaker aging [10] with recording spanning over 16 years and used during the 2010 evaluation edition.

While the primary task of the NIST SRE has always been speaker verification, various tasks derived from this one have been proposed: multi-speaker detection, which can be seen as a simplified version of speaker diarization [11, 12], speaker tracking (speaker detection as a function of time in a multi-speaker recording) or

---

[1]https://www.nist.gov/itl/iad/mig/speaker-recognition
[2]although some 16 kHz signals have been distributed as reference over the years.

semi-open set verification in 2012 [13, 14]. Human-assisted speaker recognition has also been part of the NIST SRE for two editions so far [15].

During the NIST SRE 2000, hard decision answers have been replaced by likelihoods that has since become a standard, but one can also notice the test of speaker detection cost with "no-decision" in 2002, which considered that the cost of providing no answer for a system can be less than the cost of a wrong answer.

For more details about the NIST SREs, the reader can refer to [8–10, 13, 16–21].

Although the NIST SREs are major events in the community, other benchmarking evaluations have been organized over the years to focus on different aspects of the task. In 2000, AHUMADA [22] addressed the task of speaker verification in spanish language. Speakers in the Wild [23] brought content from the Internet and re-introduced the task of speaker segmentation in the evaluation campaigns where it has been absent for a few years. In 2019, the VoxCeleb [24, 25] challenge focuses on data from the web to evaluate systems in both supervised and self-supervised speaker verification. Focusing on distant speech processing, the challenge "VOICES from a distance" addressed the challenge of reverberated speech [26]. Short duration and text dependency has been the target of the RedDot [27] and the short-duration speaker verification challenge [28]. Both challenges were evaluating speaker verification performance for different use-cases of text-dependent speaker recognition or different durations.

6.2.2   Evaluation criteria

Speaker verification is a binary classification task designed to discriminate between pairs of speech samples uttered by the same speaker and pairs of speech samples uttered by two distinct speakers. The comparison of two speech samples is referred to as a trial and a trial considering two speech segments uttered by the sample speaker is a *target* trial while a trial considering speech samples uttered by two different speakers is a *non-target* trial. When performing a trial, a verification system is expected to return to the user a binary decision describing whether the two speech samples have been uttered by the same speaker (*target* trial, hypothesis h_0) or by two different speakers (*non-target* trial, hypothesis h_1).

Such a perfect system does not exist and automatic systems are thus required to return verification scores related to their confidence in hypotheses h_0 or h_1 being true. By convention, this verification score is a scalar value and the more confident the system, the higher the verification score.

Different uses of those verification scores are possible as they can be used to produce the aforementioned binary hard decision or to compare the results of different trials in a more complex use-case.

6.2.2.1   Evaluating a system producing hard decisions

To produce a binary hard decision, automatic systems do usually compare their verification score, s (a scalar real value) to a fixed decision threshold θ. Again, the convention is that the systems validate hypothesis h_0 when $s > \theta$ and invalidate this hypothesis otherwise. Such classification of two types of trials (i.e., *target* and *non-target*) results in four cases:

- **True positive** a *target* trial is classified as such.
- **True negative** a *non-target* trial is classified as such.
- **False positive** a *non-target* trial is classified as *target*, also referred to as false alarm or Type I error.
- **False negative** a *target* trial is classified as *non-target*, also referred to as miss, false rejection or Type II error.

One can notice that for a given verification score, the binary hard decision will depend on the threshold value. Supposing that a perfect system exists, there will be a threshold that verifies $\max(s_{non\text{-}target}) < \theta < \min(s_{target})$ but in reality, distributions of *target* and *non-target* scores overlap and such a threshold cannot be found.

The reality of evaluating a speaker verification system: Evaluating the discriminatory power of a verification system corresponds to measuring its capacity to correctly classify *target* and *non-target* trials, expressed by the percentage of errors made by the system. More specifically, two error rates are measured: the false alarm rate (FAR), the number of false positives on the number of *non-target* trials; and the miss rate or false rejection rate (FRR), the number of false negatives over the number of *target* trials. The total error rate (TER) can be computed as the sum of false alarm rate and false reject rate (TER = FAR + FRR) but is not very informative as it does not indicate whether errors are due to false alarm or false reject. This is one of the reasons why the community prefers the equal error rate (EER) that provides information about the two types of errors in a single value, with EER = FAR = FRR. Indeed, the ratio between FAR and FRR is a function of the decision threshold θ. Increasing θ increases the FRR, while reducing FAR and the EER is the value of both FAR and FRR when the value of θ makes them equal. Since it is an error measure, the lower the EER, the better the system performance.

A more complete view of the system performance can be visualized on a detection error trade-off (DET) curve that represents couples (FAR, FRR) for each value of the decision threshold θ. DET curve is often preferred to the receiver operating characteristic curve in the speaker verification community as the scale of its axes provides a better resolution for low FRR and FAR values and that the shape of the DET curve provides an insight on the Gaussianity of the *target* and *non-target* score distributions. The DET curve is a straight line when both score distributions are Gaussian, its slope is determined by the standard deviation of the *target* and *non-target* score distributions whereas its distance to (0,0) is determined by the difference between the mean of the two distributions.

Evaluating speaker verification for a given application: Although EER and DET curves are good descriptors of an automatic system discrimination power, they might not be sufficient to apply speaker verification systems to real use-cases. Depending on the application, the cost of a false alarm may not be the same as the cost of a false reject. Additionally, the *a priori* probability of a *non-target* trial can be lower than the probability of a *target*. For example, in banking applications, the cost of a false alarm will be high while the number of frauds will be way lower than the number of customer accesses. On the contrary, when evaluating speaker verification systems, the combination of speech samples from a set of speakers allows to simulate a large number of *non-target* trials but only a limited number of *target* trials.

Considering this, the NIST proposed a cost detection function (DCF) that defines the cost associated to false acceptance and false reject rates as follows:

$$C_{\text{DET}} = C_{\text{FR}} \times P_{\text{target}} \times P_{\text{FR|target}} + C_{\text{FA}} \times P_{\text{non-target}} \times P_{\text{FA|non-target}} \tag{6.1}$$

In this equation, C_{FR} and C_{FA} are the costs of a false reject and false alarm, respectively, and P_{target}, the *a priori* probablity of a target trial (with $P_{\text{non-target}} = 1 - P_{\text{target}}$ depend on the application while $P_{\text{FR|target}}$ and $P_{\text{FA|non-target}}$ are system-dependent).

C_{DET} is a function of the decision threshold with its optimal value $\min C_{\text{DET}}$. The major difficulty of deploying a speaker verification system consists of setting the threshold θ such that $C_{\text{DET}} = \min C_{\text{DET}}$, a problem related to the score calibration question.

6.2.2.2 Evaluating the goodness of verification scores

The previous section described ways of measuring the discrimination power of a speaker verification system. We concluded this section by introducing the difference between C_{DET}, the cost of errors made by the system, and $\min C_{\text{DET}}$, the optimal (minimum) cost that can be obtained by adjusting the decision threshold θ.

This difference between the actual cost and the optimal one raises the question of the score meaning. So far, we have only considered the task of trial classification without considering a possible comparison between scores and by implicitly assuming that the decision threshold can be optimized, which supposes a good knowledge of the system behavior.

Calibration aims at transforming a verification score into a proper likelihood ratio, i.e., the likelihood of the two recordings in the trial given the hypothesis that they come from the same speaker divided by the likelihood of the two recordings given the hypothesis that they come from different speakers. Compared to a verification score, the likelihood ratio is an interpretable value that can be used directly in some application or be converted to binary decision via Bayes rule.

Calibration process usually converts scores to likelihood ratios, thanks to trainable parameters [29]. Those parameters are trained on a set of trials with corresponding scores in order to optimize an objective function that measures the quality of the resulting likelihood ratios. Calibration suffers from the same drawbacks as others data-driven methods and the training trials must be representative of the real trial conditions in order to guarantee the quality of the resulting likelihood ratios.

Given a score verification s, the log-likelihood ratio can be defined as

$$l(s) = \frac{P(s|\text{target})}{P(s|\text{non-target})} \tag{6.2}$$

and the Bayes decision allows to minimize the risk by accepting or rejecting the trial considering the Bayes decision threshold η, which is defined as

$$\eta = \log \frac{C_{FA}}{C_{FR}} - (P(\text{target})) \tag{6.3}$$

where C_{FA} and C_{FR} are the cost of false alarm and false reject, respectively, and $P(\text{target})$ is the target prior.

The difficulty of score calibration consists of estimating $P(s|\text{target})$ and $P(s|\text{non-target})$ as those distributions depend on the recording conditions of both the enrollment and the test samples. In controlled conditions such as benchmarking evaluation or controlled commercial applications, the estimation of those distribution is manageable but it becomes much more complicated in real cases. The task is especially difficult for forensic applications where recording conditions might not be well known and reproducible due to various factors including cooperation of the speaker.

Standard calibration methods are based on distribution estimation or matching [30–34]. Generative approaches have been proposed to use Gaussian distribution to model score distributions [35–38], but score distributions are usually more complex than simple distributions and resulting calibration can be very poor. Kernel density functions can also be used for this use [35, 39] as well as parametric distributions [40, 41]. Unsupervised generative models have been used in [42] and finally other distributions in [43]. The most standard approach in speaker recognition is the transformation of score via logistic regression that has been used for score calibration as well as score fusion [30, 31, 33, 44, 45]. Logisitic regression offers the advantage of a simple approach that estimates an affine transformation of the input dataset of scores in order to optimize an objective function.

Other approaches consider that each trial has to be considered in its uniqueness and that calibration parameters have to be trial-dependent [2, 46–48].

To measure the quality of score calibration, the *log likelihood ratio cost*, C_{llr}, has been proposed in [31]. C_{llr} is a scalar measure that has been used extensively [49–52] and is defined as follows:

$$C_{llr} = \frac{1}{N_1} \sum_{i_1} \log_2 \left(1 + \frac{1}{LR_{i_1}}\right) + \frac{1}{N_2} \sum_{j_2} \log_2 \left(1 + \frac{1}{LR_{j_2}}\right) \tag{6.4}$$

where indices i_1 and j_2 denote summing over the likelihood ratio values of the simulated cases where each hypothesis, *target* or *non-target*, is true. One advantage of the C_{llr} measure is that it can be decomposed as follows:

$$C_{llr} = C_{llr}^{\min} + C_{llr}^{cal} \tag{6.5}$$

where $C_{llr}^{\min}$ is the discrimination cost of the likelihood ratio that measures the discrimination power of the system while C_{llr}^{cal} represents the calibration cost of the system.

To conclude this section, we would like to underline the fact that in any cases, a special caution must be paid to the calibration of scores when evaluating speaker recognition systems as the usability of a system depends on the good calibration of its scores. For this reason, the question of generating an output for every trial can be raised when considering recording conditions that might not allow a proper calibration of the score as discussed by the authors of [46, 53].

6.2.3 Statistical significance

Automatic speaker verification systems are evaluated over a number of trials that are made out of speech segments recorded from numerous speakers in various conditions. A rigorous evaluation of the systems would require to measure the statistical significance of the performance indicators such as EER, min/act-DCF or C_{llr} [54]. However, this question is seldom considered in the community and most results are provided without any confidence interval of any other statistical test of significance. Traditional statistical tests are not adapted to the task and the community lacks standard tools to assess whether performance variations across systems and datasets are significant.

One difficulty of the task comes from the statistical dependency of the trials. Indeed, the number of trials is usually augmented by making use of multiple recordings from each speaker in order to compensate for the limited number of speakers. In this context, the assumption of trial independence does not hold anymore. Estimating the impact of trial dependency is a difficult issue and the very few studies on this topic are related to the impact of the multiple use of speakers [55]. It would be greatly beneficial to the community if those analyses could be extended to other factors that affect the trial independence assumption (phonetic content, language, noise, channel, etc.).

6.2.4 Specific evaluation aspects

In the previous paragraphs we described how to evaluate the core speaker recognition engine (the discriminating part of the voice biometrics solution) and the score normalization or calibration module. Both aspects are present in about all the application scenarios of voice biometrics. But some other aspects of this area are not yet covered and need specific evaluation solutions:

- Robustness against spoofing and spoofing detection. We have already introduced this point in the introductory paragraph of this chapter, highlighting the growing need to assess spoofing aspects for voice biometrics applications. Evaluating the robustness against spoofing attacks is difficult because, as in the area of viruses/malware, new attacks may appear every day. Furthermore, spreading a lot of information about how a system reacts against spoofing attacks is dangerous and evaluation protocols must take this into account. There is still too few standardization efforts for spoofing robustness evaluation and too few available resources. The main initiative is the ASVspoof challenge suite [7, 56–58].
- Forensic applications. There are several kinds of forensic applications of voice biometrics [3]. If intelligence-oriented applications are somehow evaluated, thanks to NIST SRE-like evaluations, forensic voice comparison differs radically to all other applications, both in terms of prior knowledge and in terms of decision. Forensic voice comparison needs specific evaluation protocols and specific research efforts are done in this area [45, 52, 59–65].
- Privacy protection. Protecting private and/or sensitive information about the customers becomes a hot topic during the past years [66]. Two actions were

launched recently in this direction: the Voice Privacy challenge [67][3] and the ISCA SPSC Special Interest Group.[4]

6.2.5   Evaluating related technologies

Speaker recognition is a technology that can be considered in various contexts depending on the constraints and possibilities available to the system developer.

So far, we only considered the simplest cases where systems perform speaker recognition by processing chunks of audio recordings including a single speaker in each recording. Those systems are usually obtained by running machine learning algorithms that extract some knowledge from a large quantity of training data before performing comparison of independent couples of audio chunks. Variations of use-cases can be explored by considering various training and evaluation use-cases.

Speaker recognition can be part of a multi-modal authentication systems, and voice modality is often associated to video due to multi-modality of speech and the nature of the modern communication devices. Recent developments of audio-visual speaker recognition have been brought by the MOBIO [68], NIST SRE 2018 [69] and VoxCeleb [70] evaluation campaigns. Despite the higher complexity of the multi-modal task, those evaluations have used standard metrics to evaluate the recognition systems and the evaluation aspect is not different from the speech-restrained task.

The speech data can also be enriched or made more complex by considering audio recordings including voices from several speakers. This case brings up different tasks including speaker detection and speaker diarization. Speaker detection can be seen as a close variation of speaker verification, where the question becomes: "Is a given speaker speaking in this recording?". As this task consists of answering a binary question, it is usually evaluated with the same protocol as speaker verification and the fact that several speakers might speak in a single audio chunk is not considered while evaluating the performance of systems. Speaker diarization differs from detection and verification as the question the systems have to answer is not binary anymore as it becomes: "Who speaks when?". This task is then usually divided into speaker segmentation and speaker clustering. The clustering can be limited to a single audio recording or it can be performed across recordings.

Traditionally, the performance of diarization systems is given in terms of diarization error rate (DER). This score will be defined as the ratio of the overall diarization error time to the sum of the durations of the segments that are assigned to a speaker label. The DER is computed as the fraction of speaker time that is not correctly attributed to its speaker. The diarization error time for each segment n is defined as

$$E(n) = T(n)[\max(N_{\text{ref}}(n), N_{\text{sys}}(n)) - N_{\text{Correct}}(n)] \tag{6.6}$$

[3]https://www.voiceprivacychallenge.org
[4]https://www.spsc-sig.org

where $T(n)$ is the duration of segment n, $N_{ref}(n)$ is the number of speakers that are present in segment n of the reference file, $N_{sys}(n)$ is the number of system speakers that are present in segment n and $N_{Correct}(n)$ is the number of reference speakers in segment n correctly assigned by the diarization system.

$$\text{DER} = \frac{\sum_{n \in \Omega} E(n)}{\sum_{n \in \Omega}(T(n)N_{ref}(n))} \tag{6.7}$$

The diarization error time includes the time that is assigned to the wrong speaker (confusion), missed speech time and false alarm speech time. We can thus see that DER includes performance related to speaker recognition but it also evaluates the task of speech activity detection which is not usually taken into account when evaluating speaker recognition. Many evaluation campaigns have been organized across years such as REPERE [71], Albayzin [72] or NIST-RT [73].

Of course it is possible to combine multi-modality and speaker diarization, and the evaluation follows the standards of speaker diarization [74] as done during the REPERE [71] evaluation.

Other tasks combine speaker recognition in more complex protocols and thus might require specific metrics to be evaluated. Unsupervised adaptation could be an example of those although no specific metric has been developed for this task. Some protocols have been proposed like the one of NIST SRE 2005.

6.3   Bias in testing

As discussed in the introductory paragraph of this chapter, voice biometrics is more sensitive than other speech applications to assessment biases. Eliminating bias is unrealistic and it is best to highlight and understand it. The objective is to design evaluation protocols in order to neutralize the detrimental effects of biases as much as possible. In the performance evaluation protocols presented in the previous sections, two tools are generally used for this purpose. First, the set of trials performed during evaluation is divided into "conditions." A condition is defined by common characteristics shared between trials, for instance, the fact that all speech extracts are about 2 minutes long and have only been uttered by male English speakers. Second, thanks to the high computing resources available, large numbers are used. For example, the main test condition for NIST SRE 2010 [18] involved 6 000 speaker enrollments, 25 000 test segments and up to 750 000 voice comparison tests (for one testing condition). The large numbers used here give a strong impression of robustness, a little too strong even since the same voice recordings are used several times (the same audio recording is used to create one or several hundred experimental trials) as well as the speakers (the same speaker plays the role of "imposter" against many enrolled speakers), which results in a loss of independence between tests.

The number of samples per speaker, the balance of this number between the speakers and the characteristics of the speaker are seldom taken into account (except for a few characteristics, such as the sex and the mother tongue of the speaker) even if their importance is already well known [55, 75–77]. Various studies have shown that voice recordings should be analyzed more finely than in terms of numbers. Doddington *et al.* [54] have shown that only a few speakers are responsible for a large part of the errors reported. The authors have shown that performance measures significantly depend on this factor. Kahn *et al.* [78] went a step further and demonstrated that speaker recognition systems model the speech sessions and not, or not only, the speech or the voice of a given speaker. Speech sessions here refer to the entire content and context of a voice recording which includes the acoustic content of the signal in terms of speech, speaker and other noises as well as transmission artifacts due to channel or compression. The experiment was built on NIST SRE 2008 evaluation database [9] and composed of voice comparison trials, represented by a pair of speech signals (enrollment, test). For each speaker, the enrollment session that allowed the speaker recognition system to make the least errors is labeled with a "best" label. Conversely, the speech session showing the maximum number of errors is labeled with a "worst" label. The EER moves from less than 5% for the enrollments with the "best" labels to more than 20% with the "worst" labels. The phonemic content of the speech excerpts is also an important factor that needs to be controlled, as demonstrated in [63, 79]. In [80], the authors showed that the way the recordings are associated to build a voice comparison trial is also important: the homogeneity measure (between the enrollment and the test excerpts) proposed in this chapter is able by itself to predict the performance level of a speaker recognition system.

6.4   Summary and propositions

Among all speech technologies, voice biometrics is certainly the one for which evaluation is the most sensitive, as illustrated in this chapter. And this assessment is especially complicated, more than for other areas, due to the lack of clear guidelines. Nevertheless, the research works presented in this chapter show that solid knowledge exists for the evaluation of "speaker verification" aspects as well as understanding the biases and dangers of such an assessment.

Considering the level achieved by the technology, we can consider that evaluation of sufficient size is missing. Evaluating the calibration module and setting decision thresholds are still open problems today, mainly due to the lack of appropriate definitions and databases, and the limited number of samples collected per speaker from the real world.

Our main guideline is close to the "need for caution" message presented in [3, 4]: the assessment of voice biometrics must be carried out taking into account the specific nature of speaker recognition and the significant risk of bias, with the related issues in terms of security or discrimination.

To take into account the nature of the identification medium, speech, and its large variability, especially in terms of intra-speaker variability, is the first recommendation we offer. As a corollary, replacing the performance measures based on a global average by a "per speaker" logic by highlighting the speaker with the highest error rate seems an obvious first step; the robustness of a security system is always the strength of the weakest link.

In order to improve the assessment at the product level, one need to better define "voice biometry" in terms of practical solutions, i.e., make clearer the difference between a user authentication that uses speech as a medium and an authentication based on the speaker voice characteristics. The development of international standards in this area represents an interesting and maybe mandatory opportunity. Several attempts are underway, as in the VoxCrim project[5] for forensic voice comparison or the FIDO alliance[6] for commercial applications.

References

[1]   Bonastre J.F., Kahn J., Rossato S., *et al. Forensic Speaker Recognition: Mirages and Reality*. Peter Lang; 2018. pp. 250–84.

[2]   Nandwana M.K., Ferrer L., McLaren M., *et al.* 'Analysis of critical metadata factors for the calibration of SPEAKER recognition systems'. *Proc Interspeech*. 2019:4325–9.

[3]   Campbell J.P., Shen W., Campbell W.M., Schwartz R., Bonastre J.-F., Matrouf D. 'Forensic SPEAKER recognition'. *IEEE Signal Processing Magazine*. 2009;26(2):95–103.

[4]   Bonastre J.F., Bimbot F., Boë L.J., *et al.* 'Person authentication by voice: A need for caution'. *Eighth European Conference on Speech Communication and Technology*; 2003.

[5]   Bonastre J.F., Matrouf D., Fredouille C. 'Artificial impostor voice transformation effects on false acceptance rates'. *Eighth Annual Conference of the International Speech Communication Association*; 2007.

[6]   Wu Z., Evans N., Kinnunen T., Yamagishi J., Alegre F., Li H. 'Spoofing and countermeasures for SPEAKER verification: a survey'. *Speech Communication*. 2015;66:130–53.

[7]   Wu Z., Kinnunen T., Evans N. ASVspoof 2015: the first automatic speaker verification spoofing and countermeasures challenge. Sixteenth Annual Conference of the International Speech Communication Association; 2015.

[8]   Greenberg C.S., Mason L.P., Sadjadi S.O., Reynolds D.A., *et al.* 'Two decades of SPEAKER recognition evaluation at the National Institute of Standards and Technology'. *Computer Speech & Language*. 2020;60(2–3):101032.

---

[5]https://voxcrim.univ-avignon.fr
[6]https://fidoalliance.org/

[9]   Martin A.F., Greenberg C.S. NIST 2008 speaker recognition evaluation: performance across telephone and room microphone channels. Tenth Annual Conference of the International Speech Communication Association; 2009.

[10] Martin A.F., Greenberg C.S. The NIST 2010 speaker recognition evaluation. Eleventh Annual Conference of the International Speech Communication Association; 2010.

[11] Anguera Miro X., Bozonnet S., Evans N., Fredouille C., Friedland G., Vinyals O. 'Speaker diarization: a review of recent research'. *IEEE Transactions on Audio, Speech, and Language Processing*. 2012;20(2):356–70.

[12] Tranter S.E., Reynolds D.A. 'An overview of automatic SPEAKER diarization systems'. *IEEE Transactions on Audio, Speech and Language Processing*. 2006;14(5):1557–65.

[13] Greenberg C.S., Stanford V.M., Martin A.F. 'The 2012 NIST speaker recognition evaluation'. INTERSPEECH; 2013. pp. 1971–5.

[14] Martin A.F., Greenberg C.S., Howard J.M. 'Effects of the new testingparadigm of the 2012 NIST speaker recognition evaluation'. Odyssey; 2014.

[15] Greenberg C.S., Martin A.F., Doddington G.R. 'Including human expertise in speaker recognition systems: report on a pilot evaluation'. 2011 IEEE International Conference on Acoustics, Speech and Signal Processing (ICASSP); IEEE; 2011. pp. 5896–9.

[16] Martin A.F., Przybocki M.A. 'The NIST speaker recognition evaluations: 1996-2001'. 2001: A Speaker Odyssey-The Speaker Recognition Workshop; 2001.

[17] Przybocki M.A., Martin A.F., Le A.N. 'NIST speaker recognition evaluation chronicles-part 2'. 2006 IEEE Odyssey-The Speaker and Language Recognition Workshop; IEEE; 2006. pp. 1–6.

[18] Greenberg C.S., Martin A.F., Barr B.N. 'Report on performance results in the NIST 2010 speaker recognition evaluation'. Twelfth Annual Conference of the International Speech Communication Association; 2011.

[19] Greenberg C.S., Bans´e D., Doddington G.R. 'The NIST 2014 speaker recognition i-vector machine learning challenge'. Odyssey: The Speaker and Language Recognition Workshop; 2014. pp. 224–30.

[20] Bansé D., Doddington G.R., Garcia-Romero D. 'Summary and initial results of the 2013-2014 speaker recognition i-vector machine learning challenge'. Fifteenth Annual Conference of the International Speech Communication Association; 2014.

[21] Sadjadi S.O., Kheyrkhah T., Tong A. 'The 2016 NIST speaker recognition evaluation'. Interspeech; 2017. pp. 1353–7.

[22] Ortega-Garcia J., Gonzalez-Rodriguez J., Marrero-Aguiar V. 'AHUMADA: a large speech corpus in Spanish for SPEAKER characterization and identification'. *Speech Communication*. 2000;31(2-3):255–64.

[23] McLaren M., Ferrer L., Castan D. 'The 2016 speakers in the wildspeaker recognition evaluation'. Interspeech; 2016. pp. 823–7.

[24] Nagrani A., Chung J.S., Xie W., *et al.* 'Voxceleb: large-scale SPEAKER verification in the wild'. *Computer Speech & Language*. 2020;60(4):101027.

[25] Perez M., Jin W., Le D., *et al.* 'VoxCeleb2: deep speaker recognition'. Interspeech; 2018. pp. 1898–902.

[26] Nandwana M.K., Lomnitz M., Richey C. 'The VOiCES from a distance challenge 2019: analysis of speaker verification results and remaining challenges'. Proc. Odyssey 2020 The Speaker and Language Recognition Workshop; 2020. pp. 165–70.

[27] Lee K.A., Larcher A., Wang G. The RedDots data collection for speaker recognition. Sixteenth Annual Conference of the International Speech Communication Association; 2015.

[28] Zeinali H., Lee K.A., Alam J., *et al. Short-duration Speaker Verification (SdSV) Challenge 2020: The Challenge Evaluation Plan. arXiv preprint arXiv:191206311*; 2019.

[29] David A. van L., Brummer N. 'The distribution of calibrated likelihood-ratios in speaker recognition'. *Annual Conference of the International Speech Communication Association (INTERSPEECH)*; 2013. pp. 1619–23.

[30] Pigeon S., Druyts P., Verlinde P. 'Applying logistic regression to the fusion of the NIST'99 1-Speaker submissions'. *Digital Signal Processing*. 2000;10(1–3):237–48.

[31] Brümmer N., Du Preez J. 'Application-independent evaluation of speaker detection'. *Computer Speech & Language*. 2006;20(2–3):230–75.

[32] Van Leeuwen D.A., Brümmer N. 'An introduction to application-independent evaluation of speaker recognition systems'. *Speaker Classification I*. Springer; 2007. pp. 330–53.

[33] Morrison G.S. 'Tutorial on logistic-regression calibration and fusion:converting a score to a likelihood ratio'. *Australian Journal of Forensic Sciences*. 2013;45(2):173–97.

[34] Brümmer N., Doddington G. *Likelihood-Ratio Calibration Using Priorweighted Proper Scoring Rules. arXiv preprint arXiv:13077981*; 2013.

[35] Meuwly D. *Reconnaissance de locuteurs en sciences forensiques: l'apport d'une approche automatique. Université de Lausanne, Faculté de droit et des sciences criminelles*; 2000.

[36] Navratil J., Ramaswamy G.N. 'The awe and mystery of t-norm'. *Eighth European Conference on Speech Communication and Technology*; 2003.

[37] Cumani S. 'Normal variance-mean mixtures for unsupervised score calibration'. Annual Conference of the International Speech Communication Association INTERSPEECH; 2019. pp. 401–5.

[38] Ferrer L., Mclaren M. 'A speaker verification backend for improved calibration performance across varying conditions'. Speaker Odyssey: The Speaker and Language Recognition Workshop; 2020. pp. 372–9.

[39] Gonzalez-Rodriguez J., Fierrez-Aguilar J., Ramos-Castro D., Ortega-Garcia J. 'Bayesian analysis of fingerprint, face and signature evidences with automatic biometric systems'. *Forensic Science International*. 2005;155(2-3):126–40.

[40] Egli Nicole M. *Interpretation of partial fingermarks using an automated fingerprint identification system. Universit´e de Lausanne, Facult´e de droit et des sciences criminelles*; 2009.

[41] Haraksim R., Ramos D., Meuwly D., Berger C.E.H. 'Measuring coherence of computer-assisted likelihood ratio methods'. *Forensic Science International.* 2015;249:123–32.

[42] Brümmer N., Garcia-Romero D. 'Generative modelling for unsupervised score calibration'. 2014 IEEE International Conference on Acoustics, Speech and Signal Processing ICASSP; IEEE; 2014. pp. 1680–4.

[43] Brümmer N., Swart A., van Leeuwen D. 'A comparison of linear and non-linear calibrations for SPEAKER recognition'. Speake Odyssey; 2014.

[44] Brummer N., Burget L., Cernocky J., *et al.* 'Fusion of heterogeneous SPEAKER recognition systems in the STBU submission for the NIST SPEAKER recognition evaluation 2006'. *IEEE Transactions on Audio, Speech, and Language Processing.* 2007;15(7):2072–84.

[45] Gonzalez-Rodriguez J., Rose P., Ramos D., *et al.* 'Emulating DNA: rigorous quantification of evidential weight in transparent and testable forensic SPEAKER recognition'. *IEEE Transactions on Audio, Speech and Language Processing.* 2007;15(7):2104–15.

[46] Ferrer L., Nandwana M.K., McLaren M., *et al.* 'Toward fail-safe SPEAKER recognition: trial-based calibration with a reject option'. *IEEE/ACM Transactions on Audio, Speech, and Language Processing.* 2018;27(1):140–53.

[47] Solewicz Y.A., Koppel M. 'Considering speech quality in speaker verification fusion'. *Ninth European Conference on Speech Communication and Technology*; 2005.

[48] Solewicz Y.A., Koppel M. 'Using post-classifiers to enhance fusion of low- and high-level SPEAKER recognition'. *IEEE Transactions on Audio, Speech and Language Processing.* 2007;15(7):2063–71.

[49] Ramos D., Franco-Pedroso J., Lozano-Diez A., Gonzalez-Rodriguez J. 'Deconstructing cross-entropy for probabilistic binary classifiers'. *Entropy.* 2018;20(3):208.

[50] Morrison G. 'Likelihood-ratio-based forensic SPEAKER comparison using parametric representations of vowel formant trajectories'. *The Journal of the Acoustical Society of America.* 2009;125(4).

[51] Vergeer P., Bolck A., Peschier L.J.C., Berger C.E.H., Hendrikse J.N. 'Likelihood ratio methods for forensic comparison of evaporated gasoline residues'. *Science & Justice.* 2014;54(6):401–11.

[52] Ramos D., Gonzalez-Rodriguez J., Zadora G., Aitken C. 'Information-theoretical assessment of the performance of likelihood ratio computation methods'. *Journal of Forensic Sciences.* 2013;58(6):1503–18.

[53] Schwartz R., Campbell J.P., Shen W. 'When to punt on SPEAKER comparison?' *The Journal of the Acoustical Society of America.* 2011;130(4):2547.

[54] Doddington G., Liggett W., Martin A. 'Sheep, goats, lambs and wolves: An analysis of individual differences in speaker recognition performance'. *The*

International Conference on Spoken Language Processing; Sydney: ICSLP; 1998.

[55] Wu J.C., Martin A.F., Greenberg C.S., Kacker R.N. 'The impact of data dependence on SPEAKER recognition evaluation'. *IEEE/ACM Transactions on Audio, Speech, and Language Processing*. 2017;25(1):5–18.

[56] Kinnunen T., Sahidullah M., Delgado H., *et al.* The ASVspoof 2017 challenge: Assessing the limits of replay spoofing attack detection. Proceedings of the Annual Conference of the International Speech Communication Association, INTERSPEECH; 2017-August. pp. 2–6.

[57] Wu Z., Yamagishi J., Kinnunen T., *et al.* 'ASVspoof: the automatic SPEAKER verification spoofing and countermeasures challenge'. *IEEE Journal of Selected Topics in Signal Processing*. 2017;11(4):588–604.

[58] Todisco M., Wang X., Vestman V. ASVspoof 2019: Future Horizons in Spoofed and Fake Audio Detection. Proc. Interspeech 2019; 2019. pp. 1008–12.

[59] van Leeuwen D.A., Martin A.F., Przybocki M.A., *et al.* 'NIST and NFI-TNO evaluations of automatic SPEAKER recognition'. *Computer Speech & Language*. 2006;20(2–3):128–58.

[60] Greenberg C.S., Martin A.F., Brandschain L., *et al.* Human assisted speaker recognition in NIST SRE10 (2010). Speaker and Language Recognition Workshop; Odyssey; 2010. pp. 180–5.

[61] Morrison G.S. 'Measuring the validity and reliability of forensic likelihood-ratio systems'. *Science & Justice*. 2011;51(3):91–8.

[62] Ajili M., Bonastre J.F., Rossetto S., *et al.* 'Inter-speaker variability in forensic voice comparison: a preliminary evaluation'. *2016 IEEE International Conference on Acoustics, Speech and Signal Processing ICASSP*. IEEE; 2016. pp. 2114–18.

[63] Ajili M., Bonastre J.F., KhederW B., *et al.* 'Phonetic content impact on forensic voice comparison'. *2016 IEEE Spoken Language Technology Workshop SLT*. IEEE; 2016. pp. 210–17.

[64] Meuwly D., Ramos D., Haraksim R. 'A guideline for the validation of likelihood ratio methods used for forensic evidence evaluation'. *Forensic Science International*. 2017;276:142–53.

[65] Chanclu A., Georgeton L., Fredouille C., *et al.* 'PTSVOX: une base de données pour la comparaison de voix dans le cadre judiciaire'. *Actes des Journées d'Études sur la Parole (JEP, 31e édition), Journées d'Études sur la Parole*; 2020. pp. 73–81.

[66] Nautsch A., Jiménez A., Treiber A., *et al.* 'Preserving privacy in SPEAKER and speech characterisation'. *Computer Speech & Language*. 2019;58(2):441–80.

[67] Tomashenko N., Srivastava B.M.L., Wang X., *et al. Introducing the Voice Privacy Initiative*. arXiv preprint arXiv:200501387; 2020.

[68] Marcel S., McCool C., Matějka P., *et al.* 'On the results of the first mobile biometry (MOBIO) face and speaker verification evaluation'. *International Conference on Pattern Recognition*. Springer; 2010. pp. 210–25.

[69]   Sadjadi S.O., Greenberg C., Singer E. 'The 2019 NIST audio-visual speaker recognition evaluation'. *Proc Speaker Odyssey (submitted)*; Tokyo, Japan; 2020.

[70]   Chung J.S., Nagrani A., Coto E., *et al*. *VoxSRC 2019: The first VoxCeleb Speaker Recognition Challenge. arXiv preprint arXiv:191202522*; 2019.

[71]   Kahn J., Galibert O., Quintard L., *et al*. 'A presentation of the REPERE challenge'. *2012 10th International Workshop on Content-Based Multimedia Indexing CBMI*. IEEE; 2012. pp. 1–6.

[72]   Castán D., Tavarez D., Lopez-Otero P., *et al*. 'Albayzín-2014 evaluation: audio segmentation and classification in broadcast news domains'. *EURASIP Journal on Audio, Speech, and Music Processing*. 2015;2015(1):33.

[73]   Garofolo J.S., Fiscus J.G., Martin A.F., *et al*. *NIST Rich Transcription 2002 Evaluation: A Preview*. LREC; 2002.

[74]   Gatica-Perez D., Lathoud G., McCowan I., *et al*. 'Audio-visual speaker tracking with importance particle filters'. *Proceedings 2003 International Conference on Image Processing (Cat. no. 03CH37429)*. 3. IEEE; 2003. pp. III–25.

[75]   Wu J.C., Martin A.F., Greenberg C.S., *et al*. 'Data dependency on measurement uncertainties in speaker recognition evaluation'. *Active and Passive Signatures III*. 8382. International Society for Optics and Photonics; 2012. p. 83820D.

[76]   Wu J.C., Martin A.F., Greenberg C.S., *et al*. 'Significance test with data dependency in speaker recognition evaluation'. *Active and Passive Signatures IV*. 8734. International Society for Optics and Photonics; 2013. p. 87340I.

[77]   Wu J.C., Martin A.F., Greenberg C.S., *et al*. 'Uncertainties of measures in speaker recognition evaluation'. *Active and Passive Signatures II*. 8040. International Society for Optics and Photonics; 2011. p. 804008.

[78]   Kahn J., Rossato S., Bonastre J.F. 'Beyond doddington menagerie, a first step towards'. *2010 IEEE International Conference on Acoustics, Speech and Signal Processing*. IEEE; 2010. pp. 4534–7.

[79]   Ajili M., Bonastre J.F., Kheder W.B., *et al*. 'Phonological content impact on wrongful convictions in forensic voice comparison context'. *2017 IEEE International Conference on Acoustics, Speech and Signal Processing ICASSP*. IEEE; 2017. pp. 2147–51.

[80]   Ajili M., Bonastre J.F., Kheder W.B. 'Homogeneity measure impact on target and non-target trials in forensic voice comparison'. *Proc Interspeech*; 2017. pp. 2844–8.

Chapter 7

Voice biometrics: how the technology is standardized

Andreas Nautsch[1,2] and Christoph Busch[1]

This chapter reports on biometric standardization projects that are relevant to speaker recognition. The main focus is placed on activities within the International Organization for Standardization (ISO) and the International Electrotechnical Commission (IEC). Both host the Joint Technical Committee (JTC1), which is composed of subcommittees (SCs). The *Biometrics* subcommittee, ISO/IEC JTC1/SC37, develops biometric standards since June 2002. Its goal is to ensure a focused and comprehensive worldwide approach for rapid development and approval of formal international biometric standards. While each biometric field (e.g., voice, face, and iris recognition) represents communities of their own with different best practices, SC37 harmonizes among the biometric communities. This encompasses, among others, a general system design, a concise biometric vocabulary, performance testing and reporting, detection of presentation attacks, data protection of biometric information, as well as interoperable interfaces and data interchange formats. This chapter discusses on de facto best practices in speaker recognition in the light of biometrics standards. Even so the first is driven by different fundamentals in performance assessment and signal processing to bridge gaps appears promising to enable non-experts to assess different biometric modalities with a common method, e.g., security risks, performance ranking, interoperability across vendors, data privacy, and consumer protection.

[1]Biometrics and Internet-Security Research Group, Hochschule Darmstadt, Germany
[2]Audio Security & Privacy Research Group, EURECOM, Biot, France

7.1   Introduction

In this chapter,[1] the focus is placed on the standardization outcomes of the *Biometrics* subcommittee and their application for speaker recognition (voice biometrics). Although voice biometrics is often a part of more complex systems, such as inter-active voice response (IVR) technology, not all voice biometric applications are IVRs. Thus, standards related to voice transmission, web-based audio dialog crea-tion, telecommunication, data coding for speech recognition, and audio data formats in general are not discussed here. First, biometrics standardization within ISO/IEC is described; second, the biometric voice data interchange format is described; then, de facto standards are discussed with ISO/IEC standards, and finally, conclusions are drawn. The goal of this chapter is to make *biometrics* standardization easier acces-sible to experts in voice biometrics and vice versa.

7.2   Biometrics standardization within ISO/IEC

International standardization in the field of information technology is driven by the JTC1 formed by the ISO and the IEC. An important part of the JTC1 is the Sub-Committee 37 (SC37), which was established in 2002. First standards developed in SC37 became available in 2005 and have found wide deployment in the meantime. More than 900 million implementations according to SC37 standards are estimated to be in the field at the time of this writing. Essential topics that are covered by SC37 include the definition of a harmonized biometric vocabulary (HBV) (ISO/IEC 2382-37) that removes contradictions in the biometric terminology [1], a har-monized definition of a system architecture (contained in ISO/IEC 39794-1) that describes the distributed subsystems, which are contained in deployed systems [7], a common programming interface BioAPI (ISO/IEC 19784-1) that supports ease of integration of sensors and SDKs [8], and also the definition of data interchange formats. SC37 has over the many years of work concentrated on the development of a data interchange standards, the so-called ISO/IEC 19794 family, followed by the ISO/IEC 39794 family.

SC37[2] is organized in six working groups (WGs):

- **WG 1:** Harmonized biometric vocabulary
- **WG 2:** Biometric technical interfaces
- **WG 3:** Biometric data interchange formats
- **WG 4:** Technical implementation of biometric systems
- **WG 5:** Biometric testing and reporting
- **WG 6:** Cross-jurisdictional and societal aspects of biometrics.

[1]Parts of this chapter are based on earlier work, in particular, [1–6].
[2]https://committee.iso.org/home/jtc1sc37

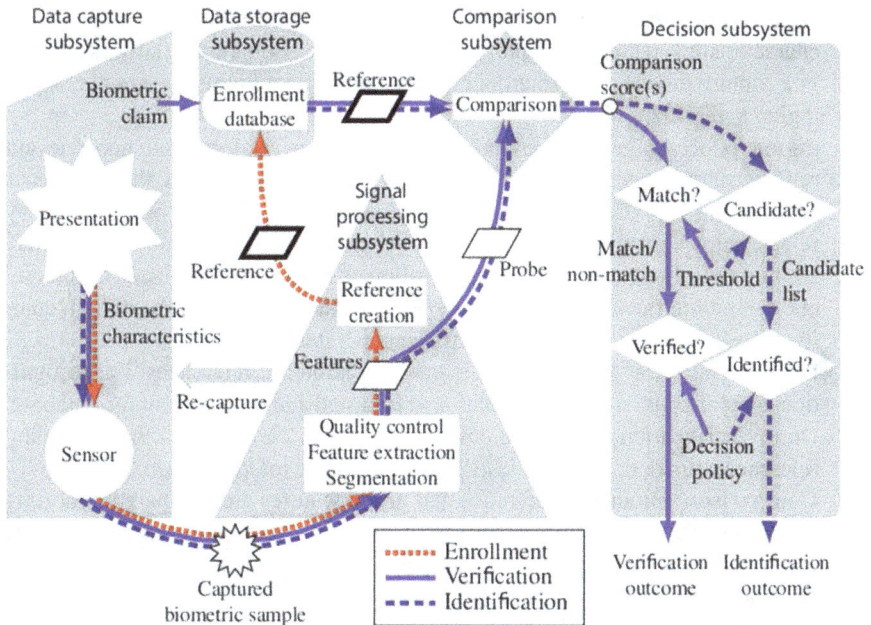

Figure 7.1   General design of a biometric system; source and shape semantics: ISO/IEC 39794-1 [7]. Reproduced from [6]

WG 1 seeks to harmonize the relevant vocabulary that is used not only across standardization projects within one WG, but also across WGs. This is especially relevant as there are different WG scopes and also different types of deliverables developed by ISO/IEC, serving a variety of purposes: (i) IS (International Standards), (ii) ITS (International Technical Specifications), (iii) TR (Technical Reports), (iv) PAS (Publicly Available Specifications), (v) IWA (International Workshop Agreements), and (vi) Guides. ISO projects are started on different tracks regarding their time frame, varying between two, three, and four years (before facing an auto-cancellation); ISO/IEC publications are revised in cycles of five years or earlier.

In the following, standards are characterized for a generalized system design, a HBV, performance testing and reporting, presentation attack detection (PAD), biometric information protection, and biometric data interchange formats (voice representations in a general framework).

7.2.1   Generalized system design

Biometric systems are composed of different subsystems. Each can potentially be supplied by different vendors: interoperability must be sustained. To process biometric data in compliance across vendors, a general system design is necessary and must be standardized. Figure 7.1 illustrates the composition of these subsystems:

- **Data capture:** Biometric capture subjects present their biological or behavioral characteristics, which are captured by biometric sensors (potentially assembled in a capture device, e.g., microphones), resulting in a biometric sample, e.g., a digital audio file.
- **Signal processing:** Samples are segmented into relevant and non-relevant regions of interest, comprising speech and non-speech, such that biometric features can be extracted. Quality control assures high system performance by rejecting low-quality samples, assuming that new samples can be recaptured.[3] High-quality features representing enrollment samples are utilized for the creation of biometric references. Probe samples that are assured to be of high quality as well are utilized in the comparison subsystem.
- **Data storage:** Databases comprise enrolled biometric references. For recognition comparisons, stored references are requested and loaded from the database. During verification, a single biometric identity is claimed and the depending reference is loaded. During identification, multiple references are loaded.
- **Comparison:** Biometric reference and probe features are compared, reporting their similarity or dissimilarity as scores, whereby verifications are one-to-one comparisons, and identifications compare a set of references against a probe, i.e., one-to-many comparisons.
- **Decision:** In a two-stage process, verification or identification outcomes are determined. First, scores are compared to thresholds, resulting in preliminary decisions or candidate lists. Second, decision policies are employed on these preliminaries from which the final recognition outcome results.

For the sake of interoperable subsystems, interfaces and data interchange formats must be harmonized. Particularly for biometric systems, an application programming interface (API) is standardized by the so-called *BioAPI*, the ISO/IEC 30106 family with the general framework [9], and programming language (Java, C#, C++) dependent specifications [10–12]. Data interchange formats, which encode biometric samples, references, and probes, are described in the next clause.

To the expert in voice biometrics, the above terminology appears stiff for describing a speaker recognition system, and the general system design appears, moreover, not necessarily well tailored to the signal processing within speaker recognition systems. The strong benefit of a general system design comes especially when interaction with non-experts in voice biometrics is demanded. For example, assessments of security risks, data privacy, and consumer protection are not carried out particularly for voice biometrics but more for biometrics in general. The use of a general system design is beyond interoperability and to avoid vendor-locks. Yet, when talking biometrics across modalities, a common language is necessary to

---

[3]Recapturing can be sustained in active capture scenarios, such as border control or mobile payments. In forensic biometrics, however, crime scene specimen cannot be recaptured, leading forensic experts to assessing the benefit of examining a specimen's sample quality in order to continue with analytic evidence reporting or refusing the specimen.

communicate concisely between experts in voice biometrics, other biometrics fields and beyond.

7.2.2 Harmonized biometric vocabulary

Literature and science specifically in a multidisciplinary community as in biometrics tend to struggle with a clear and non-contradictory use and understanding of its terms. Thus, ISO/IEC has undertaken significant efforts to develop a HBV [1] that contains terms and definitions useful also in the context of discussions about presentation attacks. The standardized terminology definition process requires that biometric concepts are always discussed in context (e.g., of one or multiple biometric subsystems) before a term and its definition for said concept can be developed. Thus, terms are defined in groups and overlap of groups (*concept clusters*) and the interdependencies of its group members necessarily lead to revision of previously found definitions. The result of this work is published as ISO/IEC 2382-37:2017 [1] and is also available online [13]. It is of interest to consider here definitions in the HBV, as they are relevant for the taxonomy and terminology defined in this chapter. The following list contains definitions of interest:

- **Biometric characteristic:** Biological and behavioral characteristic of an individual from which distinguishing, repeatable biometric features can be extracted for the purpose of biometric recognition (37.01.02)
- **Biometric sample:** Analog or digital representation of biometric characteristics prior to biometric feature extraction (37.03.21)
- **Biometric feature:** Numbers or labels extracted from biometric samples and used for comparison (37.03.11)
- **Biometric reference:** One or more stored biometric samples, biometric templates, or biometric models attributed to a biometric data subject and used as the object for biometric comparison (37.03.16)
- **Biometric probe:** Biometric sample or biometric feature set input to an algorithm for use as the subject of biometric comparison to a biometric reference(s) (37.03.14)
- **Biometric capture subject:** Individual who is the subject of a biometric capture process (37.07.03)
- **Biometric capture process:** Collecting or attempting to collect a signal(s) from a biometric characteristic, or a representation(s) of a biometric characteristic(s), and converting the signal(s) to a captured biometric sample set (37.05.02)
- **Comparison:** Estimation, calculation, or measurement of similarity or dissimilarity between biometric probe(s) and biometric reference(s) (37.05.07)
- **One-to-one comparison:** Process in which biometric probe(s) from another biometric data subject is compared to biometric reference(s) from one biometric data subject to produce a comparison score (37.05.10)
- **One-to-many comparison:** Process in which a biometric probe(s) of one biometric data subject is compared against the biometric references of more than one biometric data subject to return a set of comparison scores (37.05.10)

- **Impostor:** Subversive biometric capture subject who attempts to being matched to someone else's biometric reference (37.07.13)
- **Identity concealer:** Subversive biometric capture subject who attempts to avoid being matched to their own biometric reference (37.07.12).

The adoption of these terms is highly recommended as it avoids ambiguities not only in definitions in academic literature but also in technical specifications as they are relevant, for instance, in a call for tender. Thereby, the HBV represents the relevant consensus across all biometric modalities, leaving modality-specific terminology to their specific definitions; e.g., the term *diarization* is distinct to voice biometrics and yet not too distributed across other modalities, thus not considered in the HBV.

To the integration of multimodal biometric systems, such harmonization simplifies the composition of, e.g., voice, face, and fingerprint subsystems within governmental or commercial use case scenarios. Eventually, a well-defined vocabulary allows the meaningful translation between (technical) biometric and, e.g., ethics and data privacy communities (such as in the legal assessment of employed privacy-preserving safeguards). When different communities come together, a vocabulary harmonization provides a starting point for efficient discussions, a formally outlined standard that emerged as consensus from the reflection of different biometric communities, such as voice biometrics. A harmonized vocabulary is not only aiding discussions within the development of SC37 projects, but moreover to meaningfully interact with other communities as well; e.g., after the publication of the EU General Data Protection Regulation (GDPR) 2016/679 [14] with its distinct vocabulary, WG 1 defined their interpretation of the legal terminology in the context of generalized, harmonized biometric systems. Having such definitions enables a transparent yet efficient interaction outside of the biometrics (standardization) community, when the definition of terminology is of utmost relevance. Therefore, the harmonized vocabulary also outlines the taxonomy needed for a description of a generalized biometric system design, based on which all SC37 WGs develop their standards.

With a generalized system design and a harmonized vocabulary equipped, biometric systems are well defined for testing and reporting of biometric recognition performance of PAD and biometric information protection. The next subclauses highlight the fundamental aspects of each.

7.2.3   *Performance testing and reporting*

In order to fairly benchmark biometric systems, the standard ISO/IEC 19795-1 defines a framework for performance testing and reporting [15]. The framework differentiates fundamental performance metrics (as error rates) at different stages of a biometric system, in particular:

- **Enrolment and acquisition:** The proportion of capture attempts of biometric characteristic, which results in a failure to capture or signal of insufficient quality, is reported as failure-to-acquire (FTA) rate, whereas the proportion of population for whom the system fails to complete the enrollment process is reported as the failure-to-enroll (FTE) rate.

- **Comparison:** As algorithm performance, the Type I error rate is the false match rate (FMR) and the Type II error rate is the false non-match rate (FNMR). The first is the proportion of completed non-mated comparison trials in which the non-mated probe and reference are falsely declared to match. The latter is the proportion of completed mated comparison trials in which mated probe and reference are falsely declared not to match. These are the dominant metrics in voice biometrics research, since FTE/FTA rates are not part of the most common test sets.

- **Transaction:** Combining the FTA and both algorithm error rates in technology evaluations, the Type I error rate is the false accept rate (FAR) and the Type II error rate is the false reject rate (FRR).

- **Technology, scenario, and operational evaluation:** The FMR and the FNMR metrics are used in technology evaluations. Combining them with the FTE and the transaction error rates, in scenario or operational evaluation, the resulting Type I error rate is the generalized false accept rate (GFAR) and the resulting Type II error rate is the generalized false reject rate (GFRR). Thereby, the scenario evaluation concerns end-to-end prototypes, and the operational evaluation is conducted for one specific population and application combination.

Error rate trade-offs are visualized by the so-called detection error trade-off (DET) plots [15] (e.g., for FNMR/FMR, FAR/FRR, and GFAR/GFRR). DETs are originally motivated in [16] by observations of the speaker recognition community that many (adaptive) score normalization techniques tend to Gaussianize the score distributions [17]. The motivation of DETs is to scale the Type I and Type II error rate axes in such a way that error trade-offs rather resemble straight lines than curves to aid the visualization for an easier separability of systems' performance. Therefore, the quantile function of the standard normal distribution is used, i.e., its inverse cumulative density function (some approximations refer to log-transformed axes).[4] The standard refers to the use of linear, log, and Gaussian axes scales as *DET plot* (this might slightly diverge from the original definition in [16]); the actual axes scaling appears less of relevance to the standard.

To denote upper and lower bounds to observed error rates, the standardized performance testing and reporting framework [15] recommends uncertainty estimates in terms of confidence intervals, particularly two rules of thumb: the *rule of 3* [19] and the *rule of 30* [20].

In biometric recognition performance testing, *zero-effort impostors* are considered, whereas attacks targeting a specific biometric reference are not considering zero-effort impostors. Attackers might produce deceiving presentation attack

---

[4]In contrast to receiver operating characteristics plots [18], DET plots can depict neither 0% nor 100% due to their underlying quantile function [16], which considers a $(-\infty, +\infty)$ space.

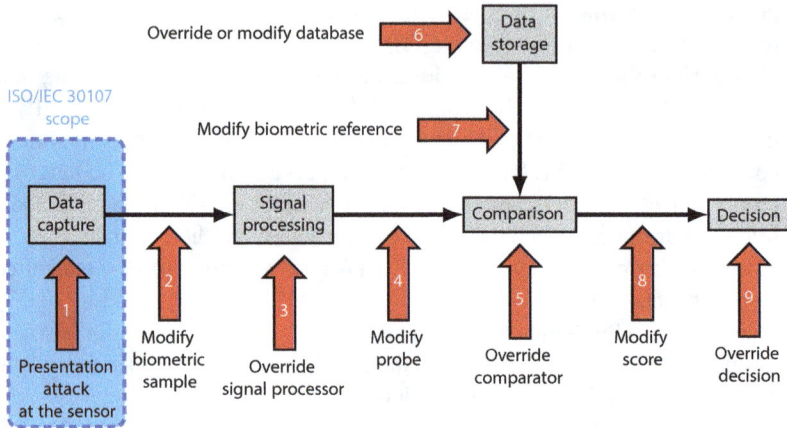

Figure 7.2   Overview on attack points in biometric systems with depicted scope of ISO/IEC 30107-1 [21]. Reproduced from [6]

instruments (PAIs) to yield higher similarity scores against the biometric reference of a specific target subject in a comparison trial [21].

7.2.4   Presentation attack detection

Presentation attacks comprise artificial, human, or other natural PAIs [21], which are presented at the capture device level to the biometric system but may also consider modified biometric samples in research evaluations [22]. Artificial PAIs comprises artifacts, such as in audio replay attacks, speech synthesis, or voice conversion. Figure 7.2 illustrates attack points to a biometric system and depicts the scope of presentation attacks according to ISO/IEC 30107-1 [21].

For detecting presentation attacks, PAD is considered an additional subsystem to the biometric system [21], see Figure 7.3: the PAD subsystem might influence the signal processing or the comparison subsystems. However, ISO/IEC 30107-1 is not restricting the system design as such. In voice PAD [23, 24], system designs consider either score fusion of comparison scores with PAD subsystem scores or a unified classifier jointly conducting biometric comparison and PAD recognition tasks.

For operational evaluations, ISO/IEC 30107-3 [22] proposes the performance assessment using detection error rates for the PAD subsystem, inspired by ISO/IEC 19795-1. For scenario and technology testing, the biometric standard family on PAD proposes to employ synthetic PAIs for offline testing. Independent of whether presentation attacks are conducted by impostors (subverting a system to be recognized as another identity, by Type I errors) or by concealers (subverting a system to remain unrecognized, by Type II errors), Type I and Type II error rates are represented by the attack presentation classification error rate (APCER) and the bona fide presentation classification error rate (BPCER). The APCER is the proportion of attack presentations using the same PAI incorrectly classified as bona fide presentations in a specific scenario [22]. The BPCER is the proportion of bona

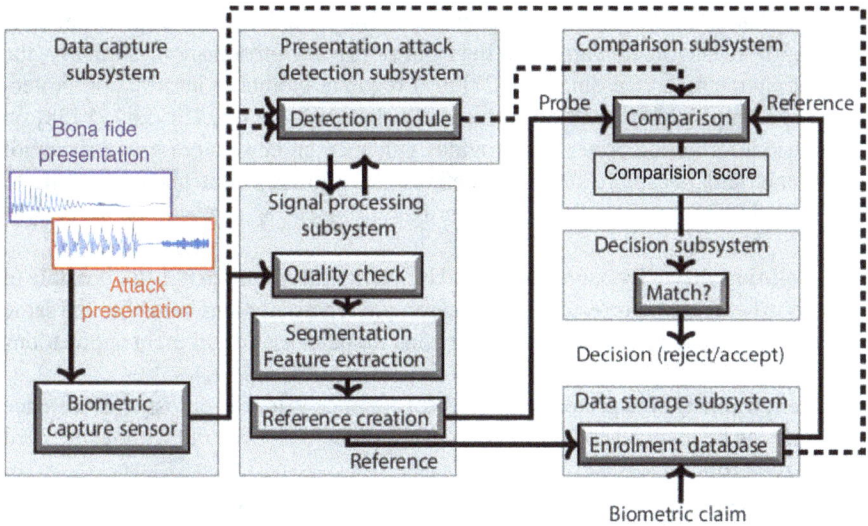

Figure 7.3   *General biometric system composition with PAD; source and shape semantics: ISO/IEC 30107-1 [21]. Reproduced from [6]*

fide presentations incorrectly classified as presentation attacks in a specific scenario [22]. For the purpose of targeting operating points enforcing security at a defined level, the standard recommends (but does not require) to report the BPCER at a 5% APCER. This working point is referred to as BPCER20. Other PAD-related metrics consider non-responses, duration efficiency, the data capture subsystem, and full system evaluation.

In contrast to the biometric standardization community, the ongoing PAD research challenge of the speaker recognition community, the ASVspoof challenge, employs a tandem decision risk [25] as a figure of merit in its 2019 edition [26]. This tandem risk is computed as a weighted error rate (which considers all cost factors relevant to the system operation regarding recognition). It is motivated from information theory and resembles the expected risk to a system operator of employing a speaker recognition system as well as a PAD subsystem.

PAD is the security perspective on biometric systems if they are used out of their purpose. Another way to subversively interact with a biometric system is to exploit sensitive data: biometric information from a privacy perspective.

7.2.5   Biometric information protection

The protection of sensitive information is important, especially in today's interconnected technology. Where several international legislations promote the protection of biometric information, SC37 complements harmonized performance testing and reporting for such.

Regarding legislation—efforts are happening globally—the European parliament enacted pioneering privacy legislation, to which today's technology needs to keep up. The 2016 EU GDPR [14] declares biometric information as personal data, i.e., highly sensitive and entitled to the right of privacy preservation. Similarly, the current payment service directive [27] also requires biometric information protection to be employed in banking services. To that end, the ISO/IEC 24745 [28] on biometric information protection provides guidance on how to preserve privacy of the biometric data capture subject. It defines the following three main properties to be fulfilled by biometric privacy safeguards:

- **Unlinkability:** Given only protected biometric information, it is not possible to say whether two protected biometric sample representations belong to the same subject. This prevents cross-comparisons for databases of different applications and ensures the privacy of the subject.
- **Renewability:** If a protected biometric reference is leaked or lost, the reference data can be revoked and renewed from the same biometric trait without the need to re-enroll.
- **Irreversibility:** Recovering biometric data from leaked protected biometric information is impossible without knowing the secret (key or algorithm) used to protect the biometric information. Restoring valid biometric features or samples is thereby prevented.

In addition to these properties, other performance metrics, such as recognition accuracy, should be preserved. The implementation of privacy safeguards faces different challenges in the public and commercial sector, however. For stand-alone systems, biometric information protection schemes might be easily implemented on top of existing systems. Nonetheless, privacy-by-design and privacy-by-default principles strongly discourage such. For multiowner systems (e.g., in the public sector) that need to provide interoperability among different owners (e.g., forensic labs in different countries), any stand-alone solution for biometric information protection is difficult to implement. Ideally, interoperability is provided through standards, which are implemented and tested for conformance.

Even if some authors argue that there is no need for biometric information protection (depending on the feature extraction) [29], sensitive information can be derived from unprotected biometric references, as has already been proved for other biometric characteristics [30, 31]. In particular, linkability of state-of-the-art speaker recognition features is demonstrated in [32] with the motivation of interchanging features among different voice biometric services. The interchange of biometric data across services is ethically addressed in [33], especially when targeting forensic scenarios. Accounting for latest data privacy legislation, biometric information needs to be protected, especially for commercial and also for forensic application scenarios with sustained precision (and also in distributed system architecture), while privacy is preserved and data is protected.

7.3 Data interchange formats for passports and beyond

Biometric data interchange must be general in its format to accommodate a plethora of representation types for each biometric modality (e.g., voice, fingerprint, face, and iris). The purpose of this section is twofold: for one, characteristic aspects of the biometric voice data interchange format are shown; for another, the placement of one modality (voice) in a general framework is showcased. The biometric voice data interchange format standard, ISO/IEC 19794-13:2018 [3], shares a generic structure and a generic header with other standards of the ISO/IEC 19794 family, a so-called *biometric data interchange record* (*BDIR*). Before explaining the format structure of BDIRs, this clause introduces parts (standards) of the ISO/IEC 19794 family and provides more information on their motivation and background.

In the conversations with speaker recognition experts, the benefit of standard interchange formats is often not seen immediately. The following three questions mimic typical arguments for the purpose of guiding readers with a speaker recognition and speech processing background on the benefits of SC37 data interchange formats.

Is it really easier for experts in voice biometrics to spend time on generic formats and then having a short description of what actually is relevant to voice data when we could just use waveform files as we are used to? **No.**

But experts in voice biometrics who are not interacting outside their realm are also not in the scope of why one needs for *standards.*

What is the benefit of a standard voice biometric data format to a standard waveform file? **It depends.**

When SC37 standards are used in technology security evaluation, e.g., such security evaluations might simply reference SC37 standards. SC37 standard-conformant interoperability would then allow for a faster accreditation of security levels. For the biometric voice data standard, it is also possible to implement the standard as an add-on to an existing solution.

So, to think beyond waveform files, as we use them, means to allow others to interact with us, especially when discussing on how they can contribute to the security measures, privacy safeguards, and interoperability among and with our voice biometric systems? **Yes.**

To engage with standardization makes voice biometrics easily accessible to those who contribute to it for a holistic solutions that are ready to meet the demands of our today's digital society whose consumers' reflection on their needs change rapidly as the technology evolves.

In the following, motivation and background are provided for encoding biometric data in harmonized format beyond voice biometrics. Then, the ISO/IEC 19794 standard family is presented (general framework and parts specific for different biometric modalities). The evolution of the format structure is described thereafter on the examples of on-card (ePassport) data storage of fingerprint and facial data (among other modalities). By consequence, the format evolved over time where interoperability, by now, needs to facilitate three generations (binary, XML, and

ASN.1-based encoding). A general structure of ISO/IEC biometric data interchange formats is shown. Within a comprehensive framework, the actual specification of *voice data* (ISO/IEC 19794 Part 13) is shown in its distinguishing characteristic. Whereas these appear abundantly obvious to the expert in voice biometrics, these specifications are neither obvious to experts in other biometric modalities and beyond biometrics.

7.3.1   Motivation and background on encoding biometric data

Biometric systems are characterized by the fact that essential functional components are usually dislocated. While the enrollment may take place as part of an employment procedure in a personal office or at a help-desk, the biometric verification often takes place at different location and time. The same holds true for forensic applications, where the enrollment is done under supervision of an officer and traces are either collected on a crime scene or extracted by forensic agencies from a video or audio recording. No matter whether the recognition system operates in verification- (one-to-one comparison) or identification (one-to-many comparison) mode, it must be capable of comparing the probe biometric data captured from the subject with the stored reference data. Applications vary in the architecture, especially with respect to the storage of the biometric reference. Some applications store the reference in a database (either centralized or decentralized), whereas other applications utilize token-based concepts like the ePassport in which subjects keep control of their personal biometric data as they decide themselves whether and when they provide the token to the controlling instance.

Given the expected complexities in system architecture, the use of open standardized formats is highly recommended as a best practice. This also relates to forensic investigations, where governmental agencies hosting biometric data in a large variety of proprietary data formats must internationally interact and exchange data ad-hoc. Any open system implementation requires the use of an interoperable, open standard that can communicate with components to be supplied from different vendors. The selection of a proprietary technology from one single vendor could add significant risk, where multiple issues (either technology or economic) could cause system failure to guarantee service. The Indian UIDAI is a good example for the benefits of standards to a large-scale biometric deployment process [34]. Furthermore, sometimes it is desired that the same biometric reference could be used in different applications: It may serve as a trusted traveler document or as ID for eGovernment applications. Applications that may be quite different in nature will require the biometric data to be encoded in one harmonized record format. Due to the nature of the different biometric characteristics being observed, an extensive series of standards is required. Some biometric systems measure biological characteristics of the individual that reflect anatomical and physiological structures of the body. Examples of these types are facial or finger characteristics. Other biometric systems measure dynamic behavioral characteristics, usually by collecting measured samples over a given time span. Examples are signature/sign data that are captured with digitizing tables or advanced pen systems or voice data that is recorded in speaker recognition systems. The ISO/IEC JTC1/SC37 series of standards known as ISO/IEC 19794 (or the 19794 family) can encode data of either type. This multipart standard includes 14 parts and covers a large variety of biometric modalities including finger, face, iris, signature, hand geometry, 3D face, voice, and DNA data. The Part 5 (face data) of the 19794 family finds application in concurrent biometric passports [35].

7.3.2   Data interchange standard ISO/IEC 19794

In order to serve the need for a modality-specific open data interchange standard, SC37 WG 3 has concentrated on the development of the ISO/IEC 19794 family, which includes currently the following 14 parts:

- Part 1: Framework
- Part 2: Finger minutiae data
- Part 3: Finger pattern spectral data
- Part 4: Finger image data
- Part 5: Face image data
- Part 6: Iris image data
- Part 7: Signature/sign time series data
- Part 8: Finger pattern skeletal data
- Part 9: Vascular image data
- Part 10: Hand geometry silhouette data
- Part 11: Signature/sign processed dynamic data
- Part 12: - void -
- Part 13: Voice data
- Part 14: DNA data
- Part 15: Palm crease image data.

The first part [2] includes relevant information that is common to all subsequent modality-specific parts such as an introduction of the layered set of SC37 standards and an illustration of a general biometric system with a description of its functional subsystems, namely the capture device, signal processing subsystem, data storage subsystem, comparison subsystem, and decision subsystem. Furthermore, this framework part illustrates the functions of a biometric system such as enrollment, verification, and identification and explains the widely used context of biometric data interchange formats in the Common Biometric Exchange Formats Framework (CBEFF) structure.

Parts 2 to 15 then detail the specification and provide modality-related data interchange formats for both image interchange and template interchange on feature level. The 19794 family gained relevance as the International Civic Aviation Organization adopted image-based representations for finger, face, and iris for storage of biometric references in electronic passports. As of today, the 19794 family found applications and interest not only in governmental and forensic scenarios but also in commercial use cases, such as for banking and payment solutions. The future of biometric data interchange formats aims at an intertranslatability between binary and human readable data interchange formats based on ASN.1 (the more recently developed 39794 family) [7].

7.3.3   Format structure

The prime purpose of a biometric reference is to represent a biometric characteristic. This representation must allow a good biometric performance when being compared to a probe sample as well as allowing a compact coding as the storage capacity for some applications may be limited. A further constraint is that the encoding format must fully support the interoperability requirements. Thus, encoding of the biometric characteristic with a two-dimensional digital representation of, e.g., a fingerprint image, face image, or iris image is a prominent format structure for many applications, which is typically realized with encoding in JPEG or JPEG2000. For voice data, contributions on sample quality are necessary.

Requirements from biometric recognition applications are quite diverse: Some applications are tuned on high biometric performance (in SC37 interpreted as low error rates) in an identification scenario. Other applications are tuned to operate with a low-capacity token (parsimonious data storage volume) in a verification scenario, where biometric capture subjects carry their biometric reference with them on a token. These tokens can be RFID cards (such as ePassports) or databases, in more general terms. Where database systems are designed (to contain biometric data, the biometric data records), the record format sub-type is the appropriate encoding, especially when bringing multiple biometric modalities together (e.g., voice and fingerprint data being sub-records of a multimodal captured sample). In other applications the token capacity may be extremely limited and thus the card format sub-type that exists in ISO/IEC 19794 for the fingerprint data formats in Parts 2, 3, and 8 is the adequate encoding. Other parts such as 19794-10, which specifies the encoding of the hand silhouette, have been designed to serve implementations that are constrained by storage space. In general the concept of compact encoding with the card format is to reduce the data size of a BDIR down to its limits. This can be achieved when necessary parameters in the metadata are fixed to standard values, which makes it obsolete to store the header information along with each individual record.

For all data interchange formats it is essential to store along with the representation of the biometric characteristic essential information (metadata) on the capturing processing and the generation of the sample. Metadata that is stored along with the biometric data (the biometric sample at any stage of processing) includes not only information such as size and resolution of the (fingerprint or face) image but also relevant data that impacted the data capturing process. Examples for such metadata are the Capture Device Type ID that identifies uniquely the device that was used for the acquisition of the biometric sample and also the impression type of a fingerprint sample, which could be a plain live-scan, a rolled live scan, non-live-scan, or stemming from a swipe sensor. For voice data, metadata includes, among others, the speech encoding and the spoken language.

For biometric systems, the quality of the biometric sample is essential information that estimates how useful the data is for future recognition tasks. Quality information is also encoded as metadata to the data interchange record. Biometric systems utilize quality for a number of different reasons. Quality is used to improve captured biometric characteristics, especially when giving rapid feedback to users to help them cooperate for better capture. Quality is utilized to improve biometric fusion for multimodal systems. Quality is measured to provide metrics for capture system maintenance, operator training, and user habituation. In general an overall assessment of the sample quality is stored on a scale from 0 to 100, while some formats allow additional local quality assessment such as the fingerprint zonal quality data or minutia quality in various fingerprint encoding standards [36]. The rationale behind this quality recording is to provide information that might weigh into a recapture decision, or to drive a failure to acquire decision. A biometric system may need to exercise quality control on biometric samples, especially enrollment, to assure strong performance, especially for identification systems. Multimodal comparison solutions can utilize quality to weigh the decisions from the various comparison subsystems to improve biometric performance. Details on how to combine and fuse different information channels can be found in the ISO/IEC technical report on multibiometric fusion [37]. A local quality assessment may also be very meaningful as environmental factors (such as different pressure, moisture, or sweat may locally degrade the image quality of a fingerprint) and thus degrade biometric performance.

According to ISO/IEC 19794-1:2011 [2] the metadata in an ISO/IEC data interchange format is subdivided in information related to the entire record, which is stored in the general header, and specific information related to one individual representation (e.g., about a whole call during an IVR call), which is stored in the representation header. The existence of multiple representations is of course dependent on the application and the respective modality used (e.g., one voice representation for each prompt in an IVR call). The general structure of ISO/IEC data interchange format standards is:

1. General header
2. Representation 1 (mandatory)
   (a) Representation header
   (b) Representation data
3. Representations 2 to N (optional)
   (a) Representation header
   (b) Representation data.

This structure was not implemented in all parts of ISO/IEC 19794 in the first generation that was issued back in the year 2005. But harmonization in this regard was achieved in the revision process of these standards, which was completed in 2011 leading to the second generation of these standards. An equivalent subdivision is in the recently developed third generation [7].

7.3.4   *ISO/IEC 19794 Part 13: voice data*

A voice data interchange format is specified by the international standard ISO/ IEC 19794-13:2018 [3], which can be used for storing, recording, and transmitting digitized acoustic human voice data (speech) assumed to be from a single speaker recorded in a single session. A general content-oriented subclause describing the voice data interchange format is followed by a subclause addressing an XML schema definition. ISO/IEC 19794-13:2018 includes vocabulary in common use by the speech and speaker recognition community, as well as terminology from other ISO standards.

The voice data interchange format follows the general structure of ISO/IEC 19794-1 [2], which allows to interchange a single voice representation or multiple voice representations (e.g., uttered during calls as prompts within IVR systems). The general header of the XML interchange format comprises information about the capture of the(se) represented biometric voice samples. Each single representation allows for three different ways to represent voice data by:

- Encapsulating conventional audio exchange formats, which encode metadata information within their binary format; these voice representations are storable as a BLOB in a 19794-13 XML file;
- Encapsulating raw acoustic data, whose metadata information is definable within the sub-header of a voice representation; raw data is storable as a BLOB in a 19794-13 XML file;
- Referring to an URL link (e.g., to a server or within a local file system), where (the binary, conventional) voice representation can be found.

The most common audio file formats are supported (e.g., wav, raw, and mp3). In this way, flexibility is sustained yet reaching a harmonization consent.

7.4   Discussion: de facto and ISO/IEC standards

De facto and ISO/IEC standards differ from one another. This clause discusses gaps between them that can be easily bridged by engaging in a communication between the speaker recognition and biometric standardization communities. Our gap analysis is grouped by the general system design, performance testing and reporting, and implementations and data interchange formats.

7.4.1   *On the general system design*

In practice, the voice biometrics community is much more close to the speech processing community than to the extended biometrics community. By consequence, the plethora of state-of-the-art voice biometric systems rely on toolkits that are well established within speech and spoken language processing, the Kaldi speech processing toolkit [38] being one of the mostly used implementations. By using such toolkits, a clear distinction of a complete voice biometric system into the above-defined subsystems is not easily drawable. Instead, the voice biometrics community distinguishes between:

- **Front end:** Captured biometric samples are processed using conventional speech features, of which the Mel-cepstral coefficients (MFCCs) are one of the most prominent low-level feature representations that incorporate a well-established acoustic model for the speech signal.
- **Back end:** Based on acoustic features, biometric features are extracted, which usually are models rather than templates (despite the fact these models have fixed-length parameter representations that can easily be confused with templates). Conventionally, the voice biometric back-end also incorporates the comparison subsystem, whose outputs are scores.
- **Score normalization and calibration:** In contrast to the SC37 perception of the word score normalization (transformation of score values to specific integer intervals), the voice biometrics community has a different interpretation. For score normalization—the removal of variance and bias to standard normal scale by enforcing the law of large numbers [17]—additional databases, the so-called cohort data, are employed [39] to carry out a statistic normalization. For score calibration—an encoding of class proportions in the score value for high-precision and low-risk decision-making—the goal is to sustain a biometric system output that has the properties of likelihood ratio (LR) scores. Score normalization and calibration can be interpreted as being a part of the biometric comparison subsystem but also as being a part of the biometric decision subsystem. Since a normalization or calibration of a score is mathematically equivalent to adaptation of a (formally defined) decision policy, LR scores are treatable from either direction (upstream, from the comparison subsystem; or downstream, from the decision subsystem). From the SC37 perspective, cohort data is rather accounted by ISO/IEC 24722:2007 by means of *characterization data*; the TR describes *multimodal and other multibiometric fusion* [37].

Taking the perspective of a voice biometrics expert (outside of biometrics standardization), the outline of generalized biometric systems diverges from de facto signal processing implementations which might appear as an unnecessary overhead since existing tool-chains are easily employable by technology experts. Taking the SC37 perspective and the perspective of non-speech technology experts, voice biometric systems as a whole appear rather complicated, and the benefits of dedicated comparison subsystems are nullified (e.g., see the above argument on proprietary technology supplied by a single vendor only; LRs are another way to achieve this). Moreover, from an SC37 perspective, the outline of generalized biometric systems allows for a divide-and-conquer strategy in testing. Bridging the gap from voice biometrics to SC37 standardization means to:

- make a clear separation between back-end (biometric) signal processing and comparison subsystems;

- outline, when score normalization and/or calibration methods belong to the decision or to the comparisons subsystem;
- harmonize the concept of intermediate biometric data representations that are inputs to front end, back end, and comparison, since voice biometric systems are not always a closed system, such as IVRs.

For bridging the gap from SC37 standardization to voice biometrics, more dialog is necessary to better understand the needs of a field. Over the past two decades, technology and research leaders in speaker recognition were barely actively involved within the due biometrics standardization processes. Nonetheless, SC37 always considered the traditional Gaussian mixture model (GMM)–universal background model (UBM) systems when developing the HBV and standards that could also impact such systems. In 2018, the SC37 committee bridged one gap toward the biometrics community by the publication of a biometric voice data interchange format, whose deployment is seamless for concurrent technology.

7.4.2 Gap analysis: performance testing and reporting

The gap of the performance testing and reporting standard in biometrics, ISO/IEC 19795-1 [15], to the state-of-the-art performance evaluation (in speaker recognition) is due to the fact that the standard has not been changed for over 15 years (when it was initially drafted). The revision of ISO/IEC 19795-1 [40] is expected to be published in 2021. The standard solely considers evaluation results in terms of error rates *as is* rather than their predictive power, the relative information one gains by using them, or even the (monetary) risk that is caused by using one biometric system over another. The most dominant examples for international research challenges are the speaker recognition evaluations (SREs) of the US National Institute of Standards and Technology (NIST) and the automatic speaker verification (ASV) anti-spoofing (ASVspoof) challenges of the academic community-driven ASVspoof consortium. Due to these (at least biannual) challenges, new data is regularly injected into the industrial and academic research community, leveraging the state of the art but also harmonizing the voice biometry community by means of speech signal processing and biometric comparison algorithms.

NIST SRE and ASVspoof editions—as well as forensic speaker recognition— make use of the Bayes theorem not only in system development (machine learning) but also in system evaluation. The following gaps of the ISO/IEC 19795-1:2006 standard [15] to the speaker recognition community are identified:

- **Proper scoring rules—the goodness of score calibration:** The standard does not include the concept of setting and defining a threshold value before any performance testing is done. By consequence, the way speaker recognition systems are tested since decades in the de facto standard way is non-standard conforming, a paradox. The goodness of score calibration informs on the forecaster accuracy of a biometric system based on its scores. As testing reports are likely to be employed for performance prediction of future decision-making, the prediction accuracy assessment of such reports is of utmost relevance to avoid misleading decisions that could be made when performance reports solely report error rates as is regardless of their impact to decision-making.

- **Identity inference for decision-making:** Bayesian (identity) inference is crucial for the purpose of making good decisions on average and, consequently, in order to assess the resulting Bayes (decision) risk as a performance metric. The standard addresses a traditional frequentist/empiricist perspective by limiting its scope to the exclusive reporting of error proportions, sparing any Bayesian perspective on performance assessment. Consequently, Bayesian perspectives on performance evaluation appear not to comply with the standard (as nonconformant) as depending methods remain unmentioned. Even if the debate on frequentist/empiricist versus Bayesian is a rather philosophical one, it is possible to state that, in an experimental observation, the frequentist's/empiricist's perspective on the current standard can only provide indirect and not direct answers on, e.g., the interval of true error rates given empirical error rates. In the case of true error rate intervals, the standard describes confidence intervals only without mentioning any of credibility intervals: a massive favoring of the frequentist/empiricist perspective over the Bayesian perspective on performance testing and reporting.

- **Information theory:** This standard focuses on particular application requirements for systems under test, with error rates serving as a proxy for an information theoretic perspective on performance. An abstraction of performance is necessary, however, when not all decision requirements are known during research and development of a system. While the standard targets performance and testing of one biometric product in one operational setting, this is only partially the case for speaker recognition applications. For the operation of an IVR, e.g., particular assumptions on classification costs and prior probabilities can be made. For forensic scenarios, though, classification costs remain unspecified, e.g., a forensic witness report cannot take into consideration how a judge or jury makes decisions. For communicating between forensic practitioners and a judge/jury, however, scores as system outputs and consequently possible threshold values need to be addressed in a more formalized way. When simulating for different priors only (since costs might remain inaccessible to an evaluator), the empirical cross-entropy (ECE) is computed [41, 42], a measure of relative information in bits. The ECE is the posterior divergence between score distributions of each class from an ideal classification outcome.

These gaps represent two fundamentally different approaches on denoting decision thresholds: the standard assesses proportions of errors initially to derive the corresponding operating point, a threshold value. Under the Bayesian paradigm, in contrast, threshold values are denoted before performance evaluation. However, for the premise of testing solely one application scenario with one report, i.e., no cross-application testing, the 19795-1:2006 standard [15] is well applicable, such as in research and development solely dedicated to automated border control (ABC) (between two specific airports). The outlined gaps need to be considered for cross-application systems, e.g., for smart home, online banking, or forensic applications.

Ideally, these gaps are resolved in the revision of the ISO/IEC 19795-1 standard and throughout other standards depending on it. Also, the presented gaps are not necessarily negative, since the concurrent version of 19795-1 presents the international consensus on performance testing and reporting for biometric recognition performance. From there, each biometric community can dive into their specific evaluation settings. For having a standard provides transparency to the communication with non-experts in biometrics, e.g., for composing muti-modal systems, recognizing individuals not only by the biometric voice.

7.4.3   *Regarding implementations and data interchange formats*

In 2015, the voice biometry standardization initiative[5] was formed. The harmonization of an intermediate-sized vector (i-vector) [41] feature extractor was promoted by a voluntary alliance of researchers, institutes, manufacturers, and organizations. A Python implementation including feature extraction parameters was proposed and made freely available. At the time of writing of this chapter, however, i-vector features in general, while still being widely used in various forms, are surpassed by x-vector features. Nonetheless, the benefit of standards is the easier interoperability between vendors, among many others.

In the larger research community within the International Speech Communication Association, a harmonization level is reached by the use of OpenSource software, foremost by *Kaldi* [38]. Since Kaldi provides freely available and computationally fast implementations of many state-of-the-art speech processing technology (also with deep learning), its data interchange formats and APIs have become a de facto standard for speech data processing. This holds especially true for the x-vector recipe, a deep learning biometric feature extractor based on Kaldi [42], which is widely adopted by the industrial and academic community. These adoptations, however, differ depending on the data used to train the feature extraction model. Derivative work further changes used programming languages (e.g., for using PyTorch), such that implementations differ not only in terms of model parameters but also in terms of programming languages. To add BioAPI interfaces potentially means effort in implementation as well as in adoptation and revision of the BioAPI standards. Also, the BioAPI family might need to be extended by a profile standard for Python.

For speech data, biometric voice probe, speech, is not captured and processed within a system like the ABC. In telephone-based IVR applications, for instance, speech data is encoded by country and vendor depending standards and transmitted over (landline) telephone cables, where the speech signal is further distorted; a distortion non-existent in ABC biometrics, where motion blur (compare speech reverberation), among others, is more of concern. This, among other reasons, led to various methods in acoustic filtering of

---

[5]https://www.voicebiometry.org/

speech signals and to the extraction of different acoustic features, which have advantages in certain (non-biometric) speech applications. Even for dominant acoustic speech features like MFCCs, parameterizations differ between almost any institution for the benefit of optimizing detection task depending (perhaps non-biometric) recognition performance. Speaker recognition is usually in use together with a much larger non-biometric speech applications. In such settings, acoustic features are extracted once for the sake of real-time human–computer interaction. One might recognize *what was said* additionally to *who spoke*. The former might have higher priority depending on the full application, in which voice biometrics is a subcomponent only. The definition of a biometric data interchange standard needs to meet the variety of existing standards of non-biometric environments in which voice biometrics operates and the SC37 framework which specifically aims at biometric applications. The application of the 19794 series aids, when biometric data of different modalities is stored together (the storage of voice representations together with, e.g., fingerprint, face, and iris representations). An add-on style integration to Kaldi-based implementations is possible (e.g., ISO/IEC 19794-13 allows for URL pointers within a file system).

7.5 Conclusion

International standards merit the harmonization of technology across vendors not only for tenders but also for the interchange between disciplines and makes the access to a technology easier. With the aim of sustaining interoperability and seeking conformance across vendors, international standards cannot satisfy latest technology advances in voice biometrics. Therefore, a consensus must be reached between technology experts, the society, industry, governmental institutions, and academia. ISO/IEC JTC1/SC37 offers a platform for this. There, biometric technology is made coherently understandable and available to communities outside of yet also in the demand of biometrics, including voice biometrics. This chapter is dedicated as a cornerstone in bridging the gaps between the voice biometrics and the biometric standardization communities.

To further bridge gaps between them, future engagement of voice technology experts within SC37 is necessary on the one hand. On the other hand, SC37 standards and projects need to reflect upon the very paradigms of how voice biometric systems are, e.g., evaluated regarding recognition and PAD performance. For the former, many steps are possible from commenting in revision and drafting of existing projects to the proposal of new standardization projects. For the latter, inclusiveness is paramount over exclusiveness: it appears odd if decades of research in voice biometrics could be simply in *nonconformance* with biometric standards. SC37 regularly expresses its needs for new input by *call-for-contributions*. One of them is on *biometric voice sample quality*. The standardization community has worked in the area of sample quality assessment in the recent years (ISO/IEC project family 29794). For the modalities fingerprint and iris standardized quality metrics exist, which are based on empirical studies. Thus, the core property of a quality metric to predict recognition performance was validated. Such standardized metrics for voice data do not exist and are an ideal subject of future work to promote the continuous harmonization between biometric modalities, including voice.

Acknowledgments

This research work has been funded by the German Federal Ministry of Education and Research and the Hessen State Ministry for Higher Education, Research and the Arts within their joint support of the National Research Center for Applied Cybersecurity ATHENE, Omilia—Conversational Intelligence, and by the VoicePersonae project funded by the French Agence Nationale de la Recherche (ANR).

References

[1]   ISO/IEC JTC1 SC37 Biometrics. *ISO/IEC 2382-37:2017 Information Technology – Vocabulary – Part 37: Biometrics*; 2017.

[2]   ISO/IEC JTC1 SC37 Biometrics. *ISO/IEC 19794-1:2011 Information Technology – Biometric Data Interchange Formats – Part 1: Framework*; 2011.

[3]   ISO/IEC JTC1 SC37 Biometrics. *ISO/IEC 19794-13:2018. Information Technology – Biometric Data Interchange Formats – Part 13: Voice Data*; 2018.

[4]   Busch C., Canon G. 'Biometric Data Interchange Format, Standardization' in Li S., Jain A. (eds.). *Encyclopedia of Biometrics*. Boston, MA: Springer; 2014. pp. 1–9.

[5]   Busch C. 'Standards for biometric presentation attack detection' in Marcel S., Nixon M.S., Fierrez J., Evans N. (eds.). *Handbook of Biometric Anti-Spoofing. Advances in Computer Vision and Pattern Recognition*. Springer, Cham; 2019. pp. 503–14. Available from https://doi.org/10.1007/978-3-319-92627-8_22.

[6]   Nautsch A. *Speaker recognition in unconstrained environments*. Technische Universität Darmstadt; 2019.

[7]   ISO/IEC JTC1 SC37 Biometrics. *ISO/IEC 39794-1:2019. Information Technology – Extensible biometric data interchange formats – Part 1: Framework*; 2019.

[8]   ISO/IEC JTC1 SC37 Biometrics. *ISO/IEC 19784-1:2006. Information Technology – Biometric Application Programming Interface – Part 1: BioAPI Specification*; 2006.

[9]   ISO/IEC JTC1 SC37 Biometrics. *ISO/IEC 30106-1:2016. Information Technology – Object oriented BioAPI – Part 1: Architecture*; 2016.

[10] ISO/IEC JTC1 SC37 Biometrics. *ISO/IEC FDIS 30106-2:2019. Information Technology – Object oriented BioAPI – Part 2: Java implementation*; 2019.

[11] ISO/IEC JTC1 SC37 Biometrics. *ISO/IEC 30106-3:2019. Information Technology – Object oriented BioAPI – Part 3: C# implementation*; 2016.

[12] ISO/IEC JTC1 SC37 Biometrics. ISO/IEC 30106-4:2019. Information Technology – Object oriented BioAPI – Part 4: C++ implementation; 2019. 13 Harmonized Biometric Vocabulary. 2019. Available from https://www.christoph-busch.de/standards.html [Accessed 2018-05-10].

[13] European Council. *Regulation 2016/679 of the European Parliament and of the Council on the protection of individuals with regard to the processing of*

personal data and on the free movement of such data General Data Protection Regulation; 2016.

[14] ISO/IEC JTC1 SC37 Biometrics. *ISO/IEC 19795-1:2006. Information Technology – Biometric Performance Testing and Reporting – Part 1: Principles and Framework*; 2006.

[15] Martin A., Doddington G., Kamm T., *et al*. 'The DET curve in assessment of detection task performance'. *Proc. Eurospeech*. 1997:1895–8.

[16] Navratil J., Ramaswamy G. 'The awe and mystery of t-norm'. *EuroSpeech. (ed.). Proc. ESCA Eur. Conf. on Speech Comm. and Tech*; 2003. pp. 2009–12.

[17] Fawcett T. 'An introduction to ROC analysis'. *Pattern Recognition Letters*. 2006;27(8):861–74.

[18] Jovanovic B., Levy P. 'A look at the rule of three'. *The American Statistican*. 1997:137–9.

[19] Doddington G.R., Przybocki M.A., Martin A.F., Reynolds D.A. 'The NIST SPEAKER recognition evaluation – overview, methodology, systems, results, perspective'. *Speech Communication*. 2000;31(31):225–54.

[20] ISO/IEC JTC1 SC37 Biometrics. *ISO/IEC 30107-1. Information Technology – Biometric presentation attack detection – Part 1: Framework*; 2016.

[21] ISO/IEC JTC1 SC37 Biometrics. *ISO/IEC 30107-3. Information Technology – Biometric presentation attack detection – Part 3: Testing and Reporting*; 2017.

[22] Sizov A., Khoury E., Kinnunen T., *et al*. 'Joint SPEAKER verification and antispoofing in the i-vector space'. *IEEE Trans on Information Forensics and Security*. 2015;10(4):821–32.

[23] Wu Z., Yamagishi J., Kinnunen T., *et al*. 'ASVspoof: the automatic SPEAKER verification spoofing and countermeasures challenge'. *IEEE Journal of Selected Topics in Signal Processing*. 2017;11(11):588–604.

[24] Kinnunen T., Kong A.L., Delgado H. 't-DCF: a detection cost function for the tandem assessment of spoofing countermeasures and automatic SPEAKER verification'. *Proc. Odyssey: The Speaker and Language Recognition Workshop*. 2018:312–9.

[25] ASVspoof Consortium. *'ASVspoof 2019: automatic SPEAKER verification spoofing and countermeasures challenge evaluation plan'*. ASVspoof Consortium; 2018.

[26] European Parliament, European Council. *Directive 2015/2366 of the European Parliament and of the Council of 25 November 2015 on payment services in the internal market*; 2015.

[27] 'ISO/IEC JTC1 SC27 Security Techniques. ISO/IEC 24745:2011'. *Information Technology – Security Techniques – Biometric Information Protection*. 2011.

[28] Vaquero C., Rodríguez P. 'On the need of template protection for voice authentication'. *INTERSPEECH. (ed.). Proc. Annual Conf. of the Intl. Speech Communication Association*; 2015. pp. 219–23.

[29] Cappelli R., Lumini A., Maltoni D. 'Fingerprint image reconstruction from standard templates'. *IEEE Transactions on Pattern Analysis and Machine Intelligence*. 2007;29(9):1489–503.

[30]   Galbally J., Ross A., Gomez-Barrero M., *et al.* 'Iris image reconstruction from binary templates: an efficient probabilistic approach based on genetic algorithms'. *Computer Vision and Image Understanding.* 2013;117(10):1512–25.

[31]   Glembek O., Matejka P., Plchot O., *et al.* 'Migrating i-vectors Between Speaker Recognition Systems Using Regression Neural Networks'. *INTERSPEECH. (ed.). Proc. Annual Conf. of the Intl. Speech Communication Association*; 2015. pp. 2327–31.

[32]   Moyakine E., Colonnello C., Butler J., *et al.* Discussion panel: SIIP and INGRESS Research Projects: Developing Effective and Sustainable Biometric Systems with a Global Reach; 2017. EAB Research Projects Conference, [Online] 34 E-AADHAAR - Unique Identification Authority of India. Available from https://www.eab.org/upload/documents/1279/08-eabr-pc2017_SIIP_INGRESS.zip?ts=1517990107115 [Accessed 2018-02-07].

[33]   'ISO/IEC JTC1 SC37 Biometricsbiometrics. *ISO/IEC 19794-5:2005. Information Technology - Biometric Data Interchange Formats - Part 5: Face Image Data*; 2005.

[34]   ISO/IEC JTC1 SC37 Biometrics. *ISO/IEC 29794-4 Information Technology - Biometric Sample Quality - Part 4: Finger image data*; 2010.

[35]   'ISO/IEC JTC1 SC37 Biometricsbiometrics. *International standards ISO/ IEC TR 24722, multimodal and other Multibiometric fusion.* International Organization for Standardisation; 2015.

[36]   Povey D., Ghoshal A., Boulianne G. 'The Kaldi speech recognition toolkit'. *Proc. IEEE 2011 Workshop on Automatic Speech Recognition and Understanding (ASRU.* 2011.

[37]   Sturim D.E., Reynolds D.A. 'Speaker Adaptive Cohort Selection for Tnorm in Text-independent Speaker Verification' in *ICASSP. (ed.). Proc. IEEE Intl. Conf. on Acoustics, Speech, and Signal Processing*; 2005. pp.. 741–4.

[38]   ISO/IEC JTC1 SC37 Biometricsbiometrics. 'ISO/IEC dis 19795-1. information technology – biometric performance testing and reporting – Part 1: principles and framework'. 2019.

[39]   Ramos D., Gonzalez-Rodrigues J. *Cross-entropy analysis of the information in forensic SPEAKER recognition.* Proc. IEEE Odyssey; 2008.

[40]   Ramos D., Franco-Pedroso J., Lozano-Diez A., Gonzalez-Rodriguez J. 'Deconstructing Cross-Entropy for probabilistic binary classifiers'. *Entropy.* 2018;20(3):208.

[41]   Dehak N., Kenny P.J., Dehak R., *et al.* 'Front-End factor analysis for SPEAKER verification'. *IEEE Transactions on Audio, Speech, and Language Processing.* 2011;19(4):788–98.

[42]   Snyder D., Garcia-Romero D., Povey D., *et al.* 'Deep Neural Network-based Speaker Embeddings for End-to-end Speaker Verification'. *SLT. (ed.). Proc. IEEE Spoken Language Technology Workshop*; 2016. pp. 165–70.

Chapter 8

Voice biometrics: perspective from the industry

Marcel Kockmann[1], Kevin Farrell[2], Daniele Colibro[2],
Claudio Vair[2], Anil Alexander[3], and Finnian Kelly[3]

Voice biometrics can now be considered a mature technology that is increasingly used in commercial and forensic applications. Successful commercial deployments have been made in financial, telecommunications, and other markets where there is a need for authentication as well as fraud management. Voice biometrics has also been used for public security applications by law enforcement agencies as well for performing forensic comparisons of voices. There have been several disruptive improvements at the core technology level in recent years that have attributed to this accelerated acceptance. These technological improvements have increased robustness to variable recording conditions and short recording durations, along with language and text independence, and have spurred on the development of effective solutions for the evolving real-world challenges in this space.

There are a number of companies, including small and medium enterprises, and large corporations, that now offer voice biometric solutions concentrating on specific sectors and use cases. This chapter includes contributions focusing on industry-related aspects of voice biometrics coming from a selection of companies that are actively deploying this technology. Each contribution has been written by a team belonging to a different company: LumenVox, Nuance Communications, and Oxford Wave Research, respectively:

1. Automated self-service password reset application
2. Testing related to commercial deployments
3. Forensic automatic speaker recognition.

[1] VoiceTrust/LumenVox, Munich, Germany
[2] Nuance Communications, United States
[3] Oxford Wave Research, United Kingdom

8.1   Automated password reset: an example of a commercial application using voice biometrics

8.1.1   Overview

Authentication has become a very important topic in recent years in consumer as well as business applications. Besides the classical knowledge-based authentication, possession- and biometric-based authentication is getting more and more popular. Those techniques can enhance both the security and the convenience of the authentication process of an individual to a commercial or business application, especially when multiple techniques are combined to a multi-factor authentication scheme.

While token-based authentication is often used in conjunction with a smartphone or other authentication device, biometric authentication is especially appealing for use cases where no special devices are available. Especially voice authentication over standard telephone lines is of great interest in scenarios where communication is usually carried out over the telephone or a telephone is available anyway.

In this section we will present a very common and popular application for voice authentication: automated self-service password reset. While many applications exist that strive to verify the identity of a speaker based on a sample of speech, they substantially differ in the way the voice biometric algorithms are used, the user experience and the system setup. We will present in detail how an enterprise-grade automated password reset solution works.

8.1.2   Introduction

Automated self-service password reset generally refers to resetting a forgotten or compromised password to a software application. While everybody knows the reset password link on websites, where a new password is sent to a registered email address, we mainly refer here to a slightly modified use case in a business environment.

Employees at companies of all scales usually have work accounts that they use to login to their desktop computers, running Windows, Mac OS, etc. One of most popular directory systems is most probably Microsoft's Active Directory (AD). Companies usually have strict security policies to have strong passwords that are complicated to remember and policies that enforce employees to change their passwords on a regular basis. While increasing the security for the company, this usually degrades the convenience for the individual user as people more often forget these passwords. Also, this leads to reduced productivity because it often takes a long time to manually reset the account password through the existing process via the IT hotline, etc.

The main motivation for using a voice-based system in this scenario is twofold. First, when an employee did forget the password to the Windows desktop, there is usually no access to the computer at all, so there is also no ability to gather a password via email. And usually the email account has the same password as the desktop login. Second, a physical phone is available in most offices, so there exists a freely

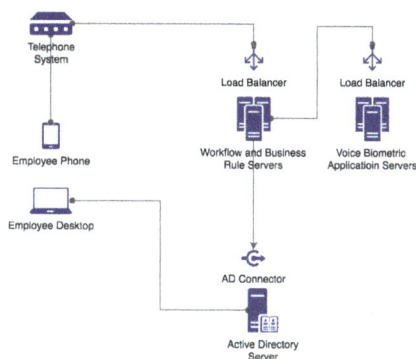

Figure 8.1   Typical architecture for a commercial automated password reset system

available device that can be used to collect the data necessary to perform a biometric authentication. So, by using a voice-based password reset system, the employee can simply pick up the phone, identify and authenticate via speech, and receive a new temporary password.

In recent years, the dissemination of mobile phone and smartphone opened new methods to reset passwords by leveraging token-based authentication. Employees can reset passwords via apps, one-time PINs (OTPs) sent via SMS or push notifications, but company phones are usually not distributed to all employees of a company, and even bring-your-own-device (BYOD) policies cannot nearly cover 100% of the employees. Similar is valid for recent trends to move directory systems such as AD to the cloud. While those systems usually offer automated password reset systems as a paid feature, many, especially large and established companies, will not put this infrastructure into a cloud-hosted environment and will often also have multiple other internal systems that they require to integrate with a customer password reset solution.

8.1.3   System architecture

A password reset system consists of several components as it needs to be tightly integrated into the customers' IT infrastructure. Figure 8.1 shows a basic architecture for an enterprise-grade password reset system.

The heart of the solution are the workflow and business rule (WBR) servers. These incorporate voice Extensible Markup Language (voice XML) or Session Initiation Protocol (SIP) handling of telephone calls and an interactive voice response (IVR) system. These servers host a workflow engine that can execute application-specific workflows. A typical, simplified workflow for a password reset application would be:

1. Call is connected, user is asked to identify via Dual-Tone Multi-Frequency (DTMF) or voice input.
2. User is asked to repeat a specific phrase.
3. On successful voice authentication, user is asked for which system the password should be reset.
4. Temporary password is spelled to the user.

It becomes apparent that those WBR servers are quite complex and incorporate a lot of different technologies. Besides, to handle the phone calls themselves and being able to model arbitrary dialogs, a full IVR system should also be fully speech enabled. Automatic speech recognition (ASR) as well as text-to-speech (TTS) technology is state of the art. One could do without though, by only using numeric DTMF input and playing back recorded phrases. As password reset systems are usually deployed in larger international enterprises, multi-language support and thus ASR and TTS in multiple languages are mandatory.

The ability to support those languages and other settings requires complex business rules, user management, administration and reporting capabilities, and further tools to support these.

Besides the core functionality of the WBR servers, the core biometric system and the backend directory services are essential components to be able to fulfill the biometric password reset functionality.

In this example we will show the simple but most common scenario: a single AD system. Many companies have access to their IT resources (employees and infrastructure) centrally managed using Microsoft's AD. Every employee and computer and their passwords are centrally registered in the companies' domain controller (DC), where password policies are set, passwords can be reset, users are disabled, etc. Employees that log into their desktop computers have to use their central AD password, which is checked against the server as soon as the computer is connected to the company network.

Programmatic access to the functionality of the AD is available through an application programming interface. The WBR servers abstract this level through the so-called connectors. For every system that is supported, a dedicated connector is available that enables the WBR system to remotely reset passwords and unlock users on the end-systems. While the figure only shows a single AD connector, a complex installation might have 5–10 different connectors to all kind of systems used in the company, like SAP systems, IBM Main Frame systems, LDAP, RSA, etc.

Besides the core functionality needed to operate an automated password reset system, a lot of supporting tools are also necessary. The directory users (employees) need to be imported into the WBR servers, key performance indicators need to be reported to the customer (either via paper/email reports or programmatic integration into a customer relationship management system), and self-service tools for employees and platform administrators are necessary.

One factor that requires increasing attention due to recent General Data Protection Regulation (GDPR) is the privacy factor. Speech samples acquired in a biometric system are a form of personal identifiable data and need to be protected accordingly. Encryption of data in motion and in rest is an absolute must, but also things like location of storage and revocation of voiceprints need to be considered. There has been recent research in the area of cancelable biometrics, but approaches like scoring in the encrypted domain are not yet easily deployable. With GDPR in effect, users at least need to give their consent to store their data and the IT systems need to provide mechanism to erase those biometric voiceprints and any personal data when the user asks for it, including backups.

8.1.4 Voice biometric system

Password reset systems normally operate in a so-called active mode, where the user directly interacts with the automated system. This is usually realized as a fully auto-mated IVR system. The so-called passive systems, where the voice biometric system just passively listens to a conversation between an agent and a customer, are more common in non-automated call center scenarios where live agents operate the incoming requests.

Text-dependent voice authentication is best suited for active systems as it requires very little audio, both in enrollment and in verification, which can be easily acquired by uttering a short phrase in the IVR dialog. Text-dependent verification consists of multiple flavors, one of them being fixed-phrase and the other being the so-called prompted.

A fixed-phrase system usually incorporates a short sentence like "My voice is my password" both for enrollment and verification. This way the lexical variabil-ity is eliminated, and best accuracy can be obtained with very little speech. Fixed-phrase systems can use a common known passphrase or multiple or even per-user phrases. A common passphrase has the main advantage that the user does not have to remember anything, because the system prompts the caller to repeat the common phrase. Further, the system can be highly optimized for this phrase, by training back-ground models specifically for this phrase, setting calibration points accordingly, and also having specific channel compensation or score normalization models. A secret user-specific passphrase does not have these benefits, but it adds an implicit second factor, a knowledge factor. Single, common fixed-phrase systems are the most widely used systems in the industry.

A very common scheme is to use three repetitions of the fixed-phrase during enrollment and a single repetition of the utterance during verification. The basic signal processing does not differentiate much from other forms of voice biometrics. The front end usually consists of standard Mel-Frequency Cepstrum Coefficients (MFCC), Linear-Frequency Cepstrum Coefficients (LFCC), or Perceptual Linear Prediction (PLP) features, and an energy- or model-based voice activity detection is used.

While the i-vector [1] paradigm and recently the x-vector [2] paradigm have been dominating the text-independent National Institute of Standards and Technology (NIST) speaker recognition evaluations [3], those models have not been very successful and popular in the text-dependent scenario. This is mainly based on the very short amounts of speech and the ability to model the speech explicitly in the text-dependent case. Time constraint models such as dynamic time warping and hidden Markov models, and channel compensation algorithm like nuisance attribute projection have been much more popular [4–6]. Also, special forms of deep neural networks (DNNs) specifically trained on a single, short utterance have been very powerful [7].

Training resources for a commercial fixed-phrase voice biometric system can be handled in several ways. One could train a generic model, even language agnostic, to be used with all deployments, or one could train a phrase-specific model, which could be for a certain language or even per customer. Our experience shows that generic, language agnostic models do not perform well in reality and it is worth going through the extra effort of training phrase models at least per language. Going through a data collection effort is usually required anyway, as at least an in-domain test set is required to set proper system calibration and operation points even with a generic background model.

A common approach to gather data is either to get it directly from the customer during a pilot phase or to acquire it from an external source. While the first is the preferred approach, it is often not possible due to privacy limitation regarding data of company employees. The latter approach creates higher up-front direct costs, as participants need to be paid to provide their speech samples. A common approach is to partner with call centers, which have many agents that are used to interact over the phone. Typically, at least 300 speakers should be engaged to provide enrollment and verification utterance over multiple days and via multiple channels. The collected data can then be used for background model training, channel compensation, cohort models, and evaluation and calibration of the system, usually by cross-validation.

Besides the voice biometric algorithms in general, there are multiple other speech and signal processing components that need to be incorporated into a commercial password reset systems. First of all, usually an automated speech recognition (ASR) system is required to initially identify the caller. Usually email addresses or employee IDs are used for that purpose, so a special ASR system tailored for, e.g., spelling is much preferred. Further, during the dialog, simple commands like yes/no or system names like Windows/SAP, etc. need to be recognized. Further, the ASR system is often leveraged to also verify that the correct authentication phrase has been spoken; this is especially important during training to ensure that no other or inconsistent phrases are enrolled. This so-called utterance verification can also be performed by special algorithm that is incorporated into the voice biometric system itself.

Further, a commercial system must provide multiple voice quality measurements to ensure that the general signal quality is of sufficient height. Usually simple measurements such as signal-to-noise ratio (SNR), amount of detected speech, clipped samples, etc. are combined with more sophisticated voice quality algorithms [8] to ensure the desired quality of the input speech.

Last but not least, commercial, fully automated systems must be protected against spoofing attacks. Especially full self-service systems in financial industries are a common target for attacks. The benefits of a fixed-phrase regarding accuracy turn into a disadvantage here. As most systems are only using a single fixed and known phrase, it is easier to try to mimic another user or try to phish recordings to assemble the requested phrase and play it back. Also, recently the advantages in TTS and voice conversion system have made such systems more vulnerable and increased the need to come up with appropriate countermeasures [9]. Usually a combination of several algorithms is deployed to detect artifacts of playback recordings, already used recordings, or synthetic speech.

8.1.5 Summary

In this section we gave an overview of the quite complex scenario of a commercial voice biometric application in the example of an automated password reset system. While the core of the application is always the voice biometric system itself striving for highest accuracy, robustness, and simplicity of deployment, it is only one of many parts in the overall system. Care must be taken to design, deploy, and maintain the overall solution to be most beneficial for the customer. Multiple generic IT systems fuse together with multiple speech processing applications like speaker recognition, speech recognition, spoofing detection, voice quality estimation, etc.

8.2 Testing of commercial voice biometric systems

8.2.1 Introduction

Speaker recognition has been gaining increased acceptance within commercial applications, such as banking. This includes applications involving self-service IVR units using text-dependent technologies to enable account access, call center applications where a customer is validated during a conversation with an agent using text-independent technology, and also fraud detection systems where text-independent models of known perpetrators are maintained in a watchlist. Successful commercial deployments of speaker recognition do require thorough testing both for setting error rate expectations and for configuring system parameters.

Speaker recognition, and specifically speaker verification, testing usually consists of evaluating data that simulates the enrollment of a user, verification of that user, and impersonation of that user. Data is collected for this assessment for as large a population as possible. Evaluations are then performed to compute the true user scores, where the enrollment and verification audio come from the same user, along with impostor scores, where the enrollment and verification audio come from different users. Given a set of true user and impostor scores, the error rates computed for the system and parameters can be configured, such as the threshold used for determining whether a score is indicative of a true user or an impostor. Within this section, we will refer to two types of errors, namely, false rejects, where the true user is mistakenly labeled as an impostor, and false accepts, where an impostor is mistakenly labeled as a true user.

Within the research community, the evaluation of speaker recognition systems has generally relied on metrics such as the equal error rate or decision cost function [10]. While these metrics can be very useful for relative comparisons of systems, they are less relevant when configuring speaker recognition systems for commercial applications. When considering commercial banking, as an example, while there may be knowledge regarding the average cost of a false accept, it is more difficult to estimate the cost of a false reject. This makes it difficult to apply cost-based metrics to these types of commercial applications where the costs themselves are unknown. Given this, most security-focused commercial applications are configured according to risk, such as the probability of having an impostor falsely accepted by the system.

As financial organizations generally have a low tolerance of risk, i.e., typically less than 1% false accept, the testing becomes more critical to ensure that thresholds are properly set. One aspect of this testing is the population size used for computing the error rates and corresponding threshold values. While large populations are required to provide reliable error rate and threshold estimates, testing more often involves smaller populations, such as hundreds of enrolled users, where audio files are reused for testing multiple models. This introduces a bias in the test results as the trials are no longer independent. Such a bias can contribute to a higher variance for the threshold estimate and the corresponding error rate. The effects of this bias will be explored in this section with respect to error rate predictions measured across the test population as well as for individual users.

8.2.1.1   Biometric testing

Error rates are computed from the number of mistakes made on a set of tests. So, for example, an estimate of the probability $\hat{p}$ of error can be computed as $\frac{e}{N}$, where e is the number of errors and N is the number of trials. Error rates are generally modeled with a binomial distribution. A binomial random variable is defined as the sum of N independent Bernoulli random variables. A Bernoulli random variable represents a single trial with a binary outcome where one class has a probability of p and the other has a probability of $1 - p$. For a large N, the binomial distribution is generally approximated with a normal distribution where the corresponding standard deviation can be computed as

$$\sigma_{\hat{p}} = \sqrt{\frac{p(1-p)}{N}} \tag{8.1}$$

Of course, when estimating a risk level for a commercial application, such as in banking, it is important to minimize the standard deviation such that the predicted error rate is aligned with that observed during system deployment. In this case, it becomes important to consider the confidence interval, i.e., the interval around an estimate that should contain the true error rate with a certain level of confidence, such as the commonly used 95% confidence interval. There are several methods for estimating the 95% confidence interval. One such approach is known as the Wald method, which is used for large values of N, where the binomial distribution is approximated as normal. The confidence interval of a probability $\hat{p}$ using the Wald method can be computed as

$$\hat{p} \pm z\sqrt{\frac{\hat{p}(1-\hat{p})}{N}} \tag{8.2}$$

where z is the quantile of a standard normal distribution for a given error rate. For example, $z = 1.96$ for 95% confidence. While it is simple, the Wald method includes several approximations and is known to have bias. A less biased, though more complicated, approach for estimating the confidence intervals is the Wilson method [11] and an exact computation of the confidence interval can be calculated using the Clopper–Pearson method [12].

Regardless of the method to estimate confidence interval, N must be selected to be as large as possible in order to minimize the potential for the true error rate to be outside a reasonable range. One guideline for selecting N is the rule-of-30 that suggests there should be enough trials to observe 30 errors [13]. So, given this criteria, a false accept rate of 1% should be computed from at least 3 000 trials. If the trials are independent, then one could expect the 95% confidence interval to be defined as roughly ±40% of the estimated error rate [14]. However, independence implies that there should not be multiple trials per speaker and audio files should only be used once. This can result in a challenging data collection given the difficulty in organizing a data collection with thousands of users.

Given this limitation, practical testing is generally performed with a biased scenario that allows multiple trials per user and reuse of audio files. A common method of testing leverages cross-comparisons where the audio of one user is used as an impostor attempt for all other users in the enrolled population. This process is repeated such that for a scenario with N enrolled users, each with one verification utterance, there will be $N * (N - 1)$ total impostor trials. While the confidence interval with this biased estimate will not be as small as that obtained with $N * (N - 1)$ unbiased trials, it will generally be better than that obtained with N trials. There has been analysis published regarding how confidence intervals with cross-comparisons [14, 15] can be estimated along with scenarios that utilize multiple samples per subject [16].

An experiment illustrating the standard deviation of the false accept rate estimate was performed using a large dataset taken from a fielded, text-dependent speaker recognition application. There are two partitions of this dataset considered in the experiment. The first partition is for a development set that consists of 13 040 male speakers, each with an enrolled model and a single verification utterance. This partition is used for estimating the threshold as corresponding to a 0.5% false accept rate. The second partition corresponds to an evaluation set with 10 751 enrolled male speakers and corresponding verification utterances. The threshold computed from the development set is used for the evaluation set to measure the actual false accept rate. Our experiment considered incremental values of N from 100 to 10 000 that include cross-comparisons using the non-match speakers. So, for example, with $N = 100$, there are $N * (N - 1) = 9\,900$ impostor trials whereas for $N = 1\,000$ there are 999 000 trials. Furthermore, for each value of N there were 100 random trials performed. We then computed the actual false accept rate and standard deviation from the 100 error rates measured using the evaluation set for each N. The results of this experiment are shown in Figure 8.2.

For this experiment, the decrease of the standard deviation of the target false accept rate for an increasing number of speakers N can be clearly observed. A rough

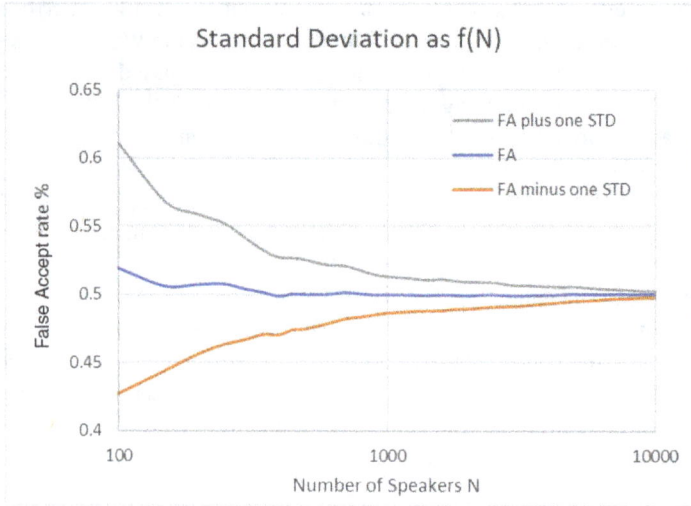

Figure 8.2   *Standard deviation of error rate estimate as a function of N. FA: false accept; STD, standard deviation.*

approximation of the 95% confidence interval could be approximated as twice the standard deviation using the Wald method.

This experiment shows the merit for improving error rates with cross-comparisons when a very large number of users are not available for estimating false accept rate. However, while the global false accept rate can be predicted with reasonable accuracy when a sufficient number of trials is available, there is still the question as to the variation in false accept exhibited at the user level. It would be ideal if all users exhibited the same false accept rates as that configured globally for the system, but as we show in the following section, this is not the case.

8.2.2   User analysis

The prior experiment considered an evaluation population of 10 751 enrolled users. For each of these enrolled users, we computed the false accept rate using the threshold computed from the development set with 10,000 users for a target false accept rate of 0.5%. In this case, each of the 10,751 enrolled users was tested with an utterance of each of the other 10,750 users and the corresponding false accept rate was recorded. A histogram of the false accept rates for these 10,751 users, including the lower and upper bounds of the 95% confidence interval, was computed and is shown in Figure 8.3.

As seen in this figure, a large number of users have twice the false accept rate of the target and there are also some users with as high as five to six times that of the target false accept rate. Such users may be considered as lambs [17] in the context of outlier users that exhibit much higher false accept rates than that of other enrolled users. Having users with large risk levels is undesirable for applications with costly

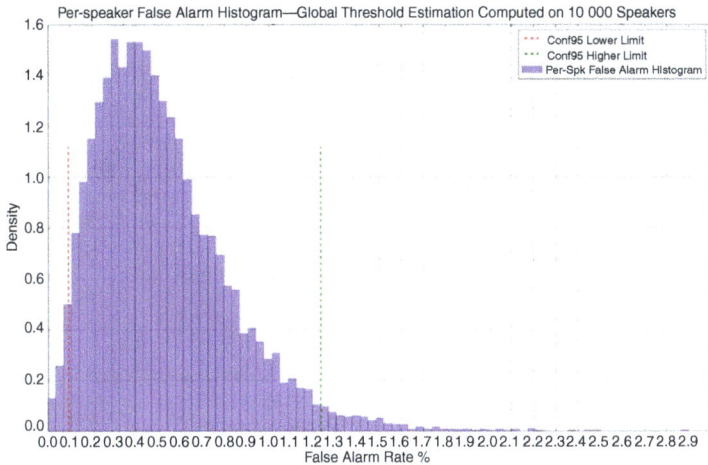

Figure 8.3 *Histogram of false accept rates across 10 000 population with global threshold*

consequences for false accepts, so a solution is required to minimize this variation in per-user false accept rate. Granted score normalization methods, such as Z-norm, have been evaluated for speaker recognition to align the impostor score distributions of enrolled users, but these methods still tend to have limitations and be only partially effective.

The method we have applied for managing the variation in per-user false accept rate is to set thresholds at the user level as opposed to a global level. In this case, we apply the non-match utterances from the development set to each model and compute model-specific thresholds to yield a specific false accept rate, i.e., 0.5% in this example. Granted, for small numbers of users this approach is inferior to that of using a global threshold, but when N is sufficiently large it does provide reduced variations to false accept rates at the user level as shown in Figure 8.4.

Here, it is seen that the 95% confidence for the user-specific threshold computation provides smaller variation than that of the global threshold computation when beyond, say, 800 users. Note that this analysis was performed for a target false accept rate of 0.5% and the crossover value of N would be higher for lower false accept rates, and correspondingly lower for higher false accept rates. The impact on the per-user false accept rates with user-specific thresholds can also be seen in the histogram shown in Figure 8.5.

When comparing the histograms for the global threshold and user-based threshold as illustrated in Figures 8.2 and 8.4, respectively, the 95% confidence interval around the 0.5% false accept target is reduced from a range of 0.1–1.25% for the global threshold to 0.3–0.8% for the user-based threshold. The extreme outliers also exhibit a large reduction to false accept rate.

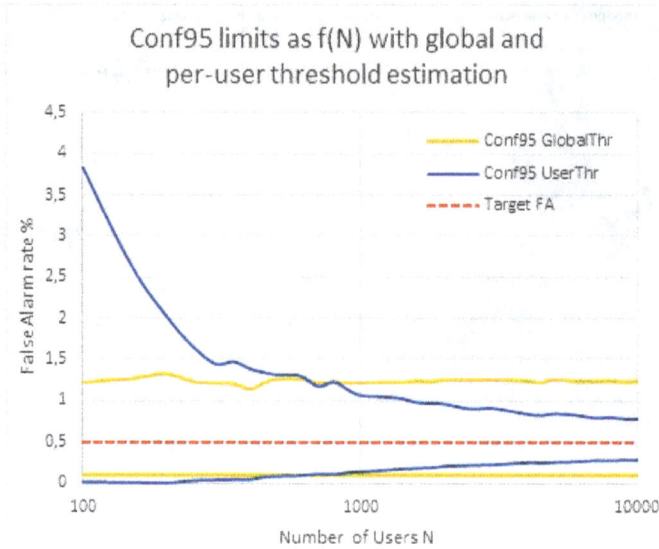

Figure 8.4   *95% Confidence interval for user false accept rate using global and per-user thresholds. FA: false accept.*

8.2.3   Summary

The analysis in this section illustrates the impact of error rate estimates based on the test population size. Cross-comparison methods are shown to be a reasonable approach to estimating the global threshold for a target false accept rate even with the dependency between trials due to the reuse of audio files. However, while this

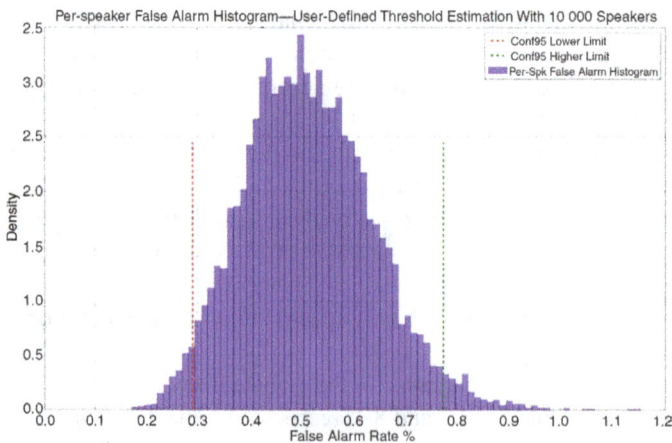

Figure 8.5   *Histogram of false accept rates across 10 000 population with per-user thresholds*

global threshold enables a good prediction of the overall false accept rate on an unseen population, there is still a large variation in false accept rate at the user level. We have addressed this by computing thresholds at the user level instead of using a global threshold. When there is a sufficient number of users available for testing, the 95% confidence interval for false accept rates at the user level can be reduced well below that achieved with a global threshold. In our experiments that have a target false accept rate of 0.5%, we found that the confidence interval would be reduced below that for the global threshold when the data from a minimum of 800 users were used to compute the per-user thresholds. When considering the full 10 000 development speakers, the global threshold exhibited a 95% confidence interval of 0.1–1.2% for a 0.5% false accept rate target. The per-user threshold approach, however, exhibited a 95% confidence interval of 0.3–0.8%, which is roughly half of that obtained with the global threshold.

8.3   Forensic speaker recognition

8.3.1   Introduction

Forensic automatic speaker recognition typically involves the comparison of speech samples using an automatic system, the estimation of the strength and validity of the comparison outcomes, and the presentation of these results to the court. Law enforcement agencies may use automatic speaker recognition for investigative purposes, which can involve, for example, the comparison of a large set of recordings with a recording of a speaker of interest, producing a shortlist of likely matches. Some of these likely matches may be further analyzed using a full forensic speaker recognition comparison process. Forensic and investigative applications of speaker recognition differ from commercial applications in several respects; it is important to highlight these differences, as they define some of the challenges faced when developing software for forensic use. Forensic applications of speaker recognition can involve completely uncontrolled recording conditions (with respect to recording environment, recording device, distance from microphone, duration, speaker profile, language and dialect, etc.). In contrast, commercial applications typically involve more constrained operating conditions, such as telephony, in-home, and in-car devices, and can be localized for a particular language or dialect area. Although recordings made in such operating conditions may be challenging, they can be controlled and prepared for in advance. In commercial applications, speakers are generally actively engaged with the service they are connecting to (such as a telephone banking application), and purposefully make an effort to be understood or identified in order to achieve a desired outcome. They may, for instance, try to speak clearly and use more formal language that is less likely to contain slang or colloquialisms. This is not typically the case in forensics, where the content and form of the speech varies widely, can be spontaneous, conversational, and often under cognitive load or duress. More generally, speech encountered in forensic cases has the potential for much intra-speaker variability in terms of the speech content, speaking style, and the emotional state of the speaker. The operators of commercial applications of speaker recognition are generally software developers or engineers. The operators in a forensic setting are primarily forensic phoneticians, for whom automatic speaker recognition is just one of a suite of tools and processes. The automatic system should therefore be compatible with their workflow and transparent in its operation.

8.3.2 *Forensic speaker recognition and the strength of evidence*

While the fundamental aim of speaker recognition is to compare two voice samples and determine the likelihood that they come from the same person or not, the circumstances in which automatic speaker recognition technology is used will dictate entirely the algorithms that can and should be applied, and how the workflow is assembled within software. In this context, law enforcement or forensic use cases have subtle but significantly different requirements to commercial verification approaches. Commercial applications are typically verification-focused and set a decision threshold in advance, based on acceptable levels of error for the specific application. In forensics, hard decisions like identification or verification are not appropriate. In forensic speaker recognition, evaluating the score obtained from the comparison of two recordings of interest in the form of a likelihood ratio (LR), and thereby quantifying the strength of evidence, is recommended [18]. The two recordings under comparison are often referred to as the questioned and reference recordings, respectively, where the questioned recording contains speech from a speaker of unknown identity and the reference recording contains speech from a speaker of known identity (although cases where the identity of the speakers on the questioned and reference recordings are unknown can occur). A LR is a means of assigning an evidential weight to the output of a speaker recognition system. Given a comparison of two speech recordings, the LR expresses the output of the system under two competing hypotheses, for instance: (1) the speech on the two recordings originates from the same speaker (i.e., the same-speaker or H0 hypothesis) and (2) the speech on the two recordings originates from different speakers (i.e., the different-speaker or H1 hypothesis) [18]. Note that the forensic expert can frame the competing hypotheses according to the specific circumstances of the case, and should explicitly state the hypothesis and any accompanying assumptions in the final report. In practice, these two hypotheses can be represented by the distribution of scores output by the system when tested with representative data. Specifically, the H0 hypothesis can be modeled as the distribution of scores produced by same-speaker comparisons and the H1 hypothesis can be modeled as the distribution of scores produced by different-speaker comparisons. The LR for a specific case, which produces a score E, can be calculated as the likelihood of E given the H0 hypothesis divided by the likelihood of E given the H1 hypothesis, as shown in Figure 8.6. The selection of representative data for generating these hypotheses is a crucial step in this process and is ultimately the responsibility of the forensic expert.

8.3.3 *The forensic expert's workflow*

Next, we present an example workflow for performing forensic casework with an automatic speaker recognition system. Although the automatic system plays a

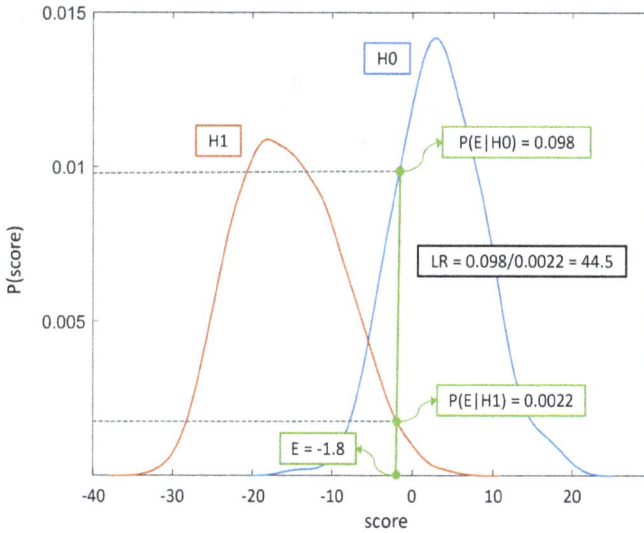

Figure 8.6   *An example of LR estimation given the same-speaker (H0) and different-speaker (H1) hypotheses and an evidential speaker recognition score* E

central and important role in this workflow, it is merely a part of this process, and the other constituent elements can affect the results and outcomes significantly. While the processing of the samples may be automatic at a signal processing level, there are many human decisions made before and after the samples are submitted to the automatic system. The following workflow example is merely illustrative, and we note that different laboratories may adopt their own standard operating procedures in handling casework.

The primary steps within a case involve:

1. Analysis of the questioned and reference recordings to assess recording quality, speech content, noise content (if any), format, sample and bit rate, etc. The expert determines whether it is possible to perform a comparison with the tools available, by verifying that the recordings satisfy minimum quality requirements (with respect to duration, SNR, clipping, etc.).

2. If minimum quality requirements are met, the questioned and reference recordings are pre-processed to ensure that speech from only one speaker is present in each recording. This is generally a manual process, although automatic speaker diarization tools can be used to assist the expert. There may be certain special cases where it is necessary to consider more than one speaker within a questioned or reference recording, but this is not typical.

3. The expert will then consider defining the relevant population and the hypotheses under consideration. For instance, if the case involves a telephone (questioned) recording that is to be compared against a police interview (reference)

recording in English, the expert would select a relevant population ideally containing speakers in both telephone and microphone conditions (similar to the police interview) in English. If ideal relevant populations cannot be sourced, it is necessary for the expert to understand the extent to which any mismatch could affect their results. At this point, the expert may also consider using additional speech data from the relevant population (if available) to adapt the system to the conditions of their case.

4. The LR (a representation of the strength of evidence) is evaluated for the case under the hypotheses defined by the expert, along with other performance metrics and graphical outputs.
5. The expert may conduct other analyses in parallel, such as a phonetic-based comparison, the outcome of which may be provided in the form of an equivalent verbal scale [18].
6. The expert will document their complete workflow and results in a final report. The report should discuss the assumptions made within the analysis, the hypotheses considered, the LR obtained, any limitations of the analysis, as well as a discussion of how the system and indeed the whole forensic analysis process have been validated within the conditions of the case.

The suite of software tools must aid a forensic expert with the workflow listed above, and do so in a reliable, repeatable way. For this reason, it is not sufficient for forensic automatic speaker recognition software to simply provide a score resulting from the comparison of two files. It is necessary for the software to support the whole speaker recognition workflow detailed above.

In recent years, as part of a broader movement in forensic science in general, there has been an increasing focus on standardization of operating procedures, validation of software tools, and expert training in forensic speaker recognition. Various organizations such as the European Network of Forensic Science Institutes (ENFSI) in Europe and the NIST-Organization of Scientific Area Committees (NIST-OSAC) in the USA have sought to establish guidelines and standards for forensic speech science. Forensic regulators across the world have sought to bring forensic speech science disciplines in line with guidelines from international standards bodies such as the ISO. As a result, forensic practitioners must ensure that their methodology and software can be validated both internally and according to external standards. Any software tools they have must therefore lend themselves to this testing and validation relatively easily; this creates a responsibility on the part of the technology provider to be transparent about how their software tools work, and how validation and calibration can be performed. The tools they develop should support the forensic expert practitioner in designing, documenting, and validating their start-to-end case workflow.

8.3.4   Technical challenges

Speaker recognition technology has been under constant evolution since its inception, with research groups across the world focused on delivering performance improvements in a multitude of different conditions (new languages, recording conditions, duration, noise, etc.). This process has been aided in part by international evaluations such as those organized by NIST. The output of this has been several generations of algorithms, along with a plethora of incremental advancements that perform well for specific conditions and under certain assumptions. The uncontrolled nature of forensic cases means that there is a wide variety in the conditions encountered, and often assumptions made about the conditions do not hold. It is therefore very important that the tools developed for forensic applications are based on algorithms and processes that are stable and thoroughly tested, and work effectively across the range of conditions that forensic practitioners encounter. Systems should lend themselves to standardization and accreditation, and the expert practitioner should be able to test the system within the conditions of the cases they encounter and understand how well the system will perform in these conditions. It is also important that these algorithms can be understood and explained by the practitioner in their reports or testimony. Systems should therefore be relatively open and use well-understood algorithms with a body of literature behind them. In this regard, the often-cited "Daubert" standard [19] for the admissibility of expert witness testimony (in the USA) refers to this criteria as whether the theory or technique is generally accepted in the scientific community and whether the technique has been subjected to peer review and publication. It is necessary that software developed for forensic use endeavors to use algorithms and methods that would satisfy such criteria. The last two decades have seen a move from Gaussian mixture model (GMM)-based speaker recognition systems to joint factor analysis, i-vector-based systems, and currently DNN-based x-vectors. While each new generation of these algorithms has undoubtedly brought an improvement in raw discrimination performance, the broader implications for use in forensic casework must also be carefully assessed. It is important for developers of systems used for forensic casework to stay up to date with the latest progress in the field, but only by embracing the stable, accepted, and general algorithms. The developers of such systems must determine which algorithms to implement, and also what settings or controls can be exposed to the user. There is a big trade-off between providing flexibility and customizable options within the software and introducing human error into the process. A further consideration is that new algorithms may have specific requirements in terms of computational power, particular operating systems, and specialized hardware such as graphic processing units. The end-users may not have access to an expensive, high-specification system, and therefore regular usage of the software should not require specialized hardware. We note that specialized functionality, such as training customized mathematical models underlying the recognition system, may only be possible with more powerful hardware. As recording conditions encountered in forensic investigation and casework vary widely, it is not conceivable that a speaker recognition system will have had sight of all possible conditions (or combinations of conditions) in advance. With the current generation of DNN-based speaker recognition systems, we observe that a useful level of discrimination is obtained for real forensic data most of the time, even when using these systems out of the box. Often, however, the user is in a position to provide representative data that can be used to adapt the system for the conditions of their case, improving discrimination performance and making the comparison scores more readily interpretable. While there have been many academic solutions proposed for system adaptation and normalization, the developers of forensic systems are faced with finding solutions that generalize effectively across many different conditions and can operate with very limited amounts of user-provided data.

8.3.4.1   Improving interpretability of scores

The numerical values of the "raw" comparison scores produced by a speaker recognition system depend on the characteristics of the test data and the configuration of the system. For example, a user may quite reasonably expect a positive score for a same-speaker comparison and a negative score for a different-speaker comparison. However, audio files recorded under different conditions can produce scores in different numerical ranges. The system may still be discriminating well between speakers, but the numerical values of the scores obtained in a new condition may not be comparable with those obtained in other conditions. The processes of score normalization and calibration bring consistency to score ranges from different conditions.

8.3.4.2   Score normalization

Score normalization is a classic approach to reducing score variability in speaker recognition. In general, score normalization adjusts the scores of a test comparison based on the comparison scores obtained from a set of reference audio files. There have been many variants proposed throughout the years (Z-norm, T-norm, etc.). An effective and flexible recent variant of score normalization is top-N symmetric score normalization (S-norm) [20]. To apply S-norm to a test score (where a test score results from the comparison of two test files), each of the test files is compared with each of the reference files, resulting in a set of reference scores for the two test files. The highest N scores (i.e. the top N) in each set of reference scores are extracted. The mean and standard deviation of the top N reference scores are calculated for each file, resulting in a set of normalization statistics for each test file. Then, two normalized scores are generated by separately applying each set of normalization statistics to the test score (normalization is applied by subtracting the mean and dividing by the standard deviation). The final S-normed score is given by the mean of the two normalized scores. In practice, we observe that top-N S-norm improves discrimination performance when there is a sufficiently large set of relevant reference files (at least 100), with most effective results in cases with general variability (e.g., differing levels of background noise) but without a systematic source of mismatch (e.g., one telephony recording vs. one microphone recording). In addition to discrimination improvements, the normalized scores are scaled and shifted into a more consistent numerical range, in a manner akin (but not equivalent to) to score calibration.

8.3.4.3 Score calibration

The process of calibration adjusts the scores so that their numerical values can be interpreted outside the context of a comparison, in specific conditions. This can be achieved by applying a linear transformation (i.e., a shifting and scaling) to the raw scores. The shifting and scaling parameters are learned in advance from a set of relevant audio files. A linear logistic regression procedure [21] is typically used to find optimal parameters given the scores obtained from these audio files. After transforming the raw scores of interest, they should lie within an expected numerical range, and the same-speaker and different-speaker scores should intersect approximately around zero. Calibration is most effective when the audio files used to learn the calibration parameters (the "calibration dataset") are representative of the conditions of the test comparison. For example, if the test comparison involves two Voice over IP (VoIP) telephony recordings of male English speakers, then a set of VoIP recordings of male English speakers would form a suitable calibration dataset. If the test comparison involves an interview recording and a VoIP recording, both containing speech from a male English speaker, then a suitable calibration dataset would consist of interview recordings of male English speakers and of VoIP recordings of male English speakers, and the scores used to learn the calibration parameters would be generated from the comparison of VoIP vs. interview recordings. If the calibration dataset is not representative of the conditions of the test comparison, then assumptions about the range and intersection point of the resulting calibrated scores do not hold. We note that scores calibrated in this way can directly be interpreted as LRs. However, we observe that the approach endorsed by ENFSI for LR estimation (i.e., the process illustrated in Figure 8.6) is preferable from an interpretation standpoint.

8.3.4.4 Condition adaptation

Since score normalization operates at the score level, its ability to improve discrimination performance is limited. A more powerful means of improving a system's discrimination performance in a specific condition is to adapt the system at the model level. There have been many academic proposals for domain adaptation operating at the speaker modeling stage (e.g., GMM, DNN) or at the speaker comparison stage (e.g., linear discriminant analysis (LDA), probabilistic LDA (PLDA)). Many of these techniques, such as PLDA parameter interpolation [22], require large volumes of adaptation data. Given the limited quantity of supplementary data typically available to the forensic practitioner, it is a major challenge to apply domain adaptation in practice. We have found that focusing on a combination of LDA and PLDA adaptation is an effective practical solution [23]. Specifically, by leveraging a weighted combination of user and system data for adapting LDA, and retraining PLDA with the adapted data, we have observed improvements in discrimination performance using small amounts of data (tens of files), particularly in cases with a systematic source of mismatch (e.g., comparison of telephony recordings with microphone recordings).

8.3.4.5   Dealing with multi-speaker recordings

Multi-speaker recordings are frequently encountered in forensic speaker recognition. A typical first step in processing such recordings is speaker diarization, which involves the labeling of speech events in a recording according to the identity of the speaker. Once diarized, speech from a speaker of interest can be extracted for speaker recognition or other analysis. Speaker diarization may be performed manually or with the assistance of diarization software. This process is time-consuming, and while it may be feasible for a small number of case recordings, it is not suitable for processing of large volumes of recordings (which may be required at the investigative stage). Consequently, the ability to bypass speaker diarization prior to speaker recognition would be a useful tool suitable for investigative data triage purposes (although not suitable for forensic LR estimation). It has been observed that the powerful discrimination performance of DNN embeddings at short durations lends itself to a sliding window approach, whereby short segments (on the order of 5 s) of a multi-speaker audio file are compared with a single-speaker audio file from a speaker of interest [24]. Applying some statistics to the resulting comparison scores can give a reliable indicator if the speaker of interest is present in the multi-speaker file.

8.3.5   Training—communication between system developers and end-users

Automatic speaker recognition forms one part of a forensic practitioners workflow. In order to use it effectively and appropriately in forensic casework, the practitioner must understand (at least at a conceptual level) the internal workings of the speaker recognition system. It is crucial that they also understand the limits of the technology and know when automatic speaker recognition should not be used. They must also have the necessary knowledge to communicate this information clearly in a case (or internal) report. There is an onus on the developers of forensic speaker recognition systems to provide this knowledge to the practitioner. Formal training, consisting of a comprehensive background introduction to the technology followed by hands-on case examples with the software, is an effective means of achieving this. In particular, working through simulated cases with data relevant to the practitioner allows specific challenges to be addressed before they occur in real casework. We note that training is not a one-way form of communication; engaging with practitioners in this way results in software that is significantly more usable and often prompts the development of new features or functionality.

8.3.6   Conclusions

The development of automatic speaker recognition software for forensic applications must address the very specific challenges and requirements of forensic casework, including the uncontrolled and variable recording conditions, along with the limited time, data, and financial budgets available to the forensic expert.

There has been a move toward accreditation and standardization in the forensic community in recent years [25], and therefore the tools and methodologies adopted by forensic practitioners must be transparent and fit within their workflow. It is no longer sufficient for manufacturers to produce speaker recognition software that simply provides a comparison result, but provide a transparent, open system that lends itself easily to validation, calibration, and adaptation to new recording conditions.

The latest generation automatic systems, particularly those based on DNNs, demonstrate good discrimination performance in forensically realistic conditions, allowing the forensic expert to obtain more reliable measures of the strength of evidence in their casework. Once part of the workflow, automatic speaker recognition allows the forensic expert to handle higher volumes of casework than traditional approaches. The ability of new systems to swiftly process large numbers of comparisons enables the expert to validate their workflow for specific conditions, such as the spoken language, duration, and quality of the recordings. At the investigative stage, automatic speaker recognition enables data triage at a scale that would not be feasible with a manual process. When integrated into the forensic workflow in this way, the new generation of automatic speaker recognition tools are poised to make a significant contribution to forensic casework.

References

[1] Dehak N., Kenny P.J., Dehak R., *et al.* 'Front-end factor analysis for speaker verification'. *IEEE Transactions on Audio, Speech, and Language Processing.* 2011;19(4):788–98.

[2] Snyder D., Garcia-Romero D., Sell G. X-vectors: robust DNN embeddings for speaker recognition. 2018 IEEE International Conference on Acoustics, Speech and Signal Processing (ICASSP); 2018. pp. 5329–33.

[3] NIST. Speaker Recognition. Available from https://www.nist.gov/itl/iad/mig/speaker-recognition [Accessed 2021/05/10].

[4] Farrell K. Speaker verification with data fusion and model adaptation. 7th International Conference on Spoken Language Processing (INTERSPEECH); Denver, Colorado (USA); 2002.

[5] Solewicz Y.A., Aronowitz H. 'Two-wire nuisance attribute projection'. Interspeech; 2009.

[6] Kenny P., Stafylakis T., Alam M.J. *Joint Factor Analysis for Text-Dependent Speaker Verification.* Joensuu, Finland: Odyssey; 2014.

[7] Heigold G., Moreno I., Bengio S., *et al. End-to-End Text-Dependent Speaker Verification* [online]. 2015. Available from https://arxiv.org/abs/1509.08062.

[8] Malfait L., Berger J., Kastner M. 'P.563—the ITU-T standard for single-ended speech quality assessment'. *IEEE Transactions on Audio, Speech and Language Processing.* 2006;14(6):1924–34.

[9] ASV. *Evaluation plan* [online]. 2019. Available from https://www.asvspoof.org/index2019.html [Accessed 05/10/2021].

[10] Przybocki M.A., Martin A.F. 'NIST speaker recognition evaluation chronicles'. 2004 IEEE Odyssey – The Speaker and Language Recognition Workshop; Toledo, Spain; 2004. pp. 12–22.

[11]   Wilson E.B. 'Probable inference, the law of succession, and statistical inference'. *Journal of the American Statistical Association*. 1927;22(158): 209–12.

[12]   Clopper C.J., Pearson E.S. 'The use of confidence or fiducial limits illustrated in the case of the binomial'. *Biometrika*. 1934;26(4):404–13.

[13]   Porter J.F. 'On the 30 error criterion' in Wayman J.L. (ed.). *National Biometric Test Center Collected Works 1997-2000*. San Jose State University, San Jose, CA, USA: National Biometric Test Center; 2000. pp. 51–6.

[14]   Wayman J.L. 'Confidence interval and test size estimation for biometric data' in Wayman J.L. (ed.). *National Biometric Test Center Collected Works 1997-2000*. San Jose State University, San Jose, CA, USA: National Biometric Test Center; 2000. pp. 91–5.

[15]   Dass S.C., Jain A.K. 'Effects of user correlation on sample size requirements' in Jain A.K., Ratha N.K. (eds.). *Biometric Technology for Human Identification II. vol. 5779*. International Society for Optics and Photonics. Orlando, FL: SPIE; 2005. pp. 226–31.

[16]   Guyon I., Makhoul J., Schwartz R., Vapnik V. 'What size test set gives good error rate estimates?' *IEEE Transactions on Pattern Analysis and Machine Intelligence*. 1998;20(1):52–64.

[17]   Doddington G., Liggett W., Martin A. *Sheep, Goats, Lambs, and Wolves: An Analysis of Individual Differences in Speaker Recognition Performance.* [ICSLP-1998] *The 5th International Conference on Spoken Language Processing*, Incorporating The 7th Australian International Speech Science and Technology Conference, Sydney Convention Centre; Sydney, Australia, 30th November - 4th December; 1998.

[18]   Drygajlo A., Jessen M., Gfrörer S., *et al*. 'Methodological Guidelines for Best Practice in Forensic Semiautomatic and Automatic Speaker Recognition: Including Guidance on the Conduct of Proficiency Testing and Collaborative Exercises. HOME/2011/ISEC/MO/4000002384'. 2015.

[19]   Daubert v. Merrell Dow Pharmaceuticals Inc. 'US Supreme Court 509 U.S. 579, 589'. 1993.

[20]   Shum S., Dehak N., Dehak R. *Unsupervised Speaker Adaptation Based on the Cosine Similarity for Text-Independent Speaker Verification*. Brno, Czech Republic: Odyssey; 2010.

[21]   Pigeon S., Druyts P., Verlinde P. 'Applying logistic regression to the fusion of the NIST'99 1-Speaker submissions'. *Digital Signal Processing*. 2000;10(1): 237–48.

[22]   Garcia-Romero D., McCree A. Supervised domain adaptation for I-vector based speaker recognition. 2014 IEEE International Conference on Acoustics, Speech and Signal Processing (ICASSP); 2014. pp. 4047–51.

[23]   Kelly F., Forth O., Kent S., Alexander A. Deep neural network based forensic automatic speaker recognition in VOCALISE using x-vectors, *Audio Engineering Society (AES) Forensics Conference 2019,* Porto, Portugal. 2019. Available from http://www.aes.org/e-lib/browse.cfm?elib=20477 [Accessed May 2021].

[24]   Snyder D., Garcia-Romero D., Sell G., *et al.* Speaker recognition for multi-speaker conversations using x-vectors. *ICASSP 2019 - 2019 IEEE International Conference on Acoustics, Speech and Signal Processing (ICASSP)*; Brighton, United Kingdom; 2019. pp. 5796–800.

[25]   European Network of Forensic Science Institutes (ENFSI). *Policy on accreditation, BRD-ACR-001*. May 2019. Available from http://enfsi.eu/wp-content/uploads/2017/06/BRD-ACR-001_Policy-on-Accreditation.pdf [Accessed March 2021].

Chapter 9

Joining forces of voice and facial biometrics: a case study in the scope of NIST SRE'19

Mohamed Amine Hmani[1], Aymen Mtibaa[1], and Dijana Petrovska Delacretaz[1]

While other chapters of this book are devoted to voice biometrics, in this chapter we present an example of joining forces of voice biometrics with other modalities. The focus of this chapter is the combination (fusion) of voice and facial biometrics, also called audiovisual biometrics. It is well known that multi-biometric systems have the advantage of improving the accuracy over single systems, providing increased security, and making spoofing attacks more difficult. Regarding voice biometrics, combining voice and facial biometrics provides specific advantages that can be exploited in various manners.

Combining voice and facial biometrics has the unique advantage that they can be acquired simultaneously, without providing an additional burden for the user. Of course, besides the microphone, a camera is needed. With the increasing use of videos, it becomes more and more natural to treat videos (with sequence of images and sound). Therefore, combining voice and facial biometrics is an interesting combination in order to provide better biometric performance while making imposture more difficult.

As an example, we can refer to the latest US National Institute of Standards and Technology (NIST) 2019 speaker recognition evaluation (SRE'19) challenge[1] that besides the usual speaker recognition challenge included for the first time an audiovisual challenge. Telecom SudParis (TSP) participated in this challenge.

This chapter provides a description of the TSP speaker and face recognition systems as well as the advantages obtained by combining voice and facial biometrics.

[1]Laboratoire SAMOVAR, Telecom SudParis, Institut Polytechnique de Paris (TSP-IPP), Palaiseau, France

9.1   Introduction to the NIST SRE'19 challenge

The SRE'19[1] [1] is part of an ongoing series of speaker recognition evaluations conducted by the NIST since 1996. They provide a common test bed that enables the research community to explore promising new ideas in speaker recognition and have a valuable impact to support the community in their development of advanced technology incorporating these ideas.

SRE'19 consisted of two separate activities. The first one was a leaderboard-style challenge using conversational telephone speech (CTS) for text-independent speaker detection. Moreover, in addition to the regular audio-only track, the SRE'19 introduced for the first time an audiovisual and visual-only tracks, denoted as multimedia track.

9.1.1   The SRE'19 CTS challenge

The task for the SRE'19 CTS challenge [1] was text-independent speaker detection. In the challenge the system has to verify whether the claimed identity is present in the test speech segment. The speech segments in the SRE'19 CTS challenge were extracted from the CallMy Net2 (CMN2) corpus comprising 8 kHz CTS in Tunisian Arabic. The CMN2 corpus was collected over the traditional public switched telephone network and the more recent voice over IP platforms outside North America.

For the SRE'19 CTS challenge, the trials are divided into two subsets: a progress subset and a test subset. The progress subset, comprising 30% of the target speakers, is used to fine-tune the parameters of the development systems, while the remaining 70% of the speakers were allocated for the test subset. The test conditions for this challenge are as follows:

1.   The speech duration of the test segments is uniformly sampled ranging from 10 to 60 seconds.
2.   Trials are conducted with test segments from same and different phone numbers as the enrollment segments.
3.   There are no cross-gender trials.

9.1.2   The SRE'19 multimedia challenge

The audiovisual data for the multimedia challenge were extracted from the unexposed portions of the Video Annotation for Speech Technology (VAST) corpus, collected by the Linguistic Data Consortium (LDC). The VAST corpus [2] contains amateur video recordings (such as video blogs) collected by the LDC from various online media hosting services. The videos vary in duration from a few seconds to several minutes and include speech spoken in English. Each video may contain audiovisual data from potentially multiple individuals who may or may not be visible in the recording. Manually produced diarization labels (i.e., speaker time marks),

---

[1]https://sre.nist.gov/

as well as key face frames and bounding boxes (that mark an individual's face in the video) were provided for both the dev set and test set enrollment videos (but not for the test videos in either set). The audio was sampled at 16 kHz.

9.1.3   SRE'19 evaluation metrics

In the NIST SRE challenges, three metrics are considered: EER, min_C, and act_C. The *EER* is the equal error rate, where the false acceptance rate (FAR) is equal to the false rejection rate (FRR).[2] act_C is the primary metric and min_C is the secondary metric.

act_C, the actual detection cost, is computed according to a basic cost model. This model is used to measure the detection performance of the submitted systems in SRE'19, which is defined as a weighted sum of false rejection and false acceptance error probabilities for some decision threshold θ. The cost function is normalized to give $C_{\mathrm{Norm}}(\theta)$, which is defined as follows:

$$C_{\mathrm{Norm}}(\theta) = P_{\mathrm{fr}}(\theta) + \beta \times P_{\mathrm{fa}}(\theta) \tag{9.1}$$

where β is defined as

$$\beta = \frac{C_{\mathrm{fa}}}{C_{\mathrm{fr}}} \times \frac{1 - P_{\mathrm{target}}}{P_{\mathrm{target}}} \tag{9.2}$$

The actual detection cost is computed from the trial scores by applying a detection threshold of $\log(\beta)$, where log denotes the natural logarithm.

For the CTS challenge, the primary cost function is computed using two thresholds. The thresholds are computed for two values of β, β_1 for $P_{\mathrm{Target_1}} = 0.01$ and β_2 for $P_{\mathrm{Target_2}} = 0.005$, where P_{target} is the *a priori* probability of the specified target speaker. We note that the prior P_{target} is a synthetic parameter used in order to reduce the multi-class problem into a binary classification problem [3, chapter 8]. The values are fixed by NIST.

In this case, the actual detection cost is the following:

$$C_{\mathrm{Primary}} = \frac{C_{\mathrm{Norm}}(\log(\beta_1)) + C_{\mathrm{Norm}}(\log(\beta_2))}{2} \tag{9.3}$$

As for the multimedia challenge, the primary cost function is defined using a single threshold. This threshold is computed for $P_{\mathrm{Target}} = 0.005$:

$$C_{\mathrm{Primary}} = C_{\mathrm{Norm}}(\log(\beta)) \tag{9.4}$$

---

[2]NIST SREs adopt the terminology "false alarm" and "miss" instead of "false acceptance" and "false rejection."

The following sections present the TSP team submission to the NIST SRE'19 for the CTS and multimedia challenges. Section 9.2 presents the speaker verification systems for the CTS and multimedia challenge. Section 9.3 presents the face recognition system for the multimedia challenge. Section 9.4 reports the audiovisual fusion for the multimedia challenge.

9.2   TSP speaker verification system for the SRE'19 evaluation

In this section, we present a brief review of the state of the art of speaker verification, followed by the description of TSP speaker verification submissions for the SRE'19 CTS and multimedia challenges.

9.2.1   A brief review of state of the art in speaker verification

Speaker verification is the process of accepting or rejecting the claimed identity of a speaker, based on his/her voice characteristics and enrolled speaker model. One of the first successful approaches for speaker verification is the Gaussian mixture model (GMM) [4]. The joint factor analysis approach [5] improves the way in which the speaker GMM is estimated by modeling speaker and channel subspaces separately. With the appearance of i-vectors [6], the speaker and the channel variabilities are defined with a single space named total variability space. From a sequence of feature vectors, e.g., Mel-frequency cepstral coefficients (MFCCs), sufficient statistics are collected and represented by Baum–Welch statistics obtained with respect to a universal background model (UBM). These statistics are converted into a low-dimensional representation known as an i-vector. After i-vectors are extracted, a probabilistic linear discriminant analysis (PLDA) model is used to produce verification scores by comparing i-vectors extracted from different speech segments.

Another i-vector framework is based on deep neural networks (DNNs) where a DNN replaces the UBM in the calculation of Baum–Welch statistics [7, 8]. Moreover, DNNs can be used to extract bottleneck characteristics, which are concatenated with MFCCs in the standard UBM/i-vector framework [9, 10].

More recently, researchers proposed an end-to-end x-vector speaker recognition systems [11]. The x-vector system uses a temporal pooling layer upon several time-delay neural networks (TDNNs) to extract speaker embedding representations. Each time-delay layer of the network gradually extracts speaker information features. Time-delay layers are followed by a statistics pooling layer designed to accumulate speaker information from the whole speech segment into a single vector, which they call an x-vector. Speaker verification systems based on x-vectors are seen as an improvement over the i-vector systems in terms of biometric performance [12], since in x-vectors, speaker features' representations are extracted from neural networks trained in a discriminant way and optimized to identify a set of speakers.

In the next subsections, we describe the TSP speaker verification system pipeline based on TDNN [11] and the extended TDNN (E-TDNN) [13], architectures that have proven their effectiveness during SRE'19.

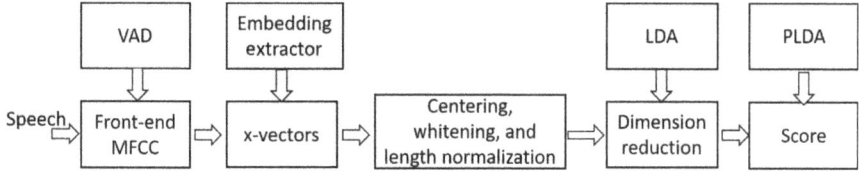

Figure 9.1   Block diagram of TSP's common speaker verification systems for NIST 2019 CTS and multimedia challenges

9.2.2   TSP speaker verification common pipeline for the SRE'19 CTS and multimedia challenges

The TSP speaker verification system for both CTS and multimedia challenges is based on TDNN embeddings called x-vectors for speaker recognition, as shown in Figure 9.1. First, preprocessing is conducted by extracting acoustic features (MFCCs) and applying Kaldi's [14] energy-based voice activity detection and normalization techniques such as mean and variance normalization. Next, a time-delay DNN, trained to classify the speakers in a set of training data, is used to map variable-length speech segments to embeddings called x-vectors. Prior to dimensionality reduction through LDA, x-vectors are centered, whitened, and unit-length normalized. Then, the comparison of pair of speaker embeddings (x-vectors) is performed using a PLDA scoring back end.

The x-vector networks are divided into three parts. First, an encoder network extracts frame-level representations from acoustic features such as MFCCs. Second, a temporal pooling layer aggregates the frame-level representations into a single vector. Last, a feed-forward classification network processes this single vector to calculate speaker class posteriors. The speaker embedding x-vector is extracted from the first affine transform after the pooling layer. Different x-vector systems are proposed in the literature characterized by different encoder architectures, pooling methods, and training objectives.

The TSP x-vectors' extractor for the CTS and multimedia challenges is based on two architectures. The first one is the TDNN, which is proposed in [11]. Its recipe is available in Kaldi toolkit [14]. The second extractor (E-TDNN) is based on an extended version of the TDNN [13], which has shown its effectiveness in the SRE18 [15].

TDNN

This embedding extractor architecture is composed of seven hidden layers and rectified linear units trained to discriminate between speakers. First, each feature frame is captured by a sequence of time-delay layers. Time-delay layers are equivalent to one-dimensional dilated convolutions. Therefore, frame-level representations at each layer aggregate information from the context of previous and next frames. The total context in this network is 15 frames. Then a temporal statistics pooling layer is employed to compute the mean and standard deviation of the TDNN output over all the frames for an input segment. The resulted segment-level representation is then fed into two fully connected layers to classify the speakers in the training set. After training, speaker embeddings are extracted from the dimensional affine component of the first fully connected layer.

E-TDNN

This architecture is an E-TDNN [13] with 12 hidden layers, where the first ten hidden layers operate at frame level while the last two operate at segment level. It was optimized to balance the trade-off between network parameters and performance by adding dense layers between the time-delay layers and extended the temporal context to 22 frames.

9.2.3   TSP speaker verification system for the SRE'19 CTS challenge

This section describes the experimental setup of TSP speaker verification systems submitted during the CTS challenge based on the fusion of TDNN and E-TDNN, the architectures commonly used during the SRE'19 evaluation. We start by presenting a summary of the data used to train the various components of the system, followed by a description of the back-end scoring.

For training the TDNN and the E-TDNN models, the SRE and Switchboard (SWB) databases were used as the training set. They correspond to data from SREs 04 to 10 (referenced here as SRE combined), Mixer 6, SWB Cellular Phases I and II, and SWB Phases I, II, and III.

Besides, data augmentation is performed to increase the amount and diversity of the available training data. The augmentation strategy is used to add four corrupted copies of the original recordings to the training list. The recordings are corrupted by employing additive noises (babble, general noise, music) from MUSAN [16] database after down-sampling the speech data to 8 kHz and reverberation that involves convolving room impulse responses (RIRs) with audio. Both MUSAN and the RIR data-sets are freely available.[3] After data augmentation, utterances under 4 s and speakers with less than eight utterances are removed from the training set.

For back-end scoring, a classifier based on PLDA is trained for the speaker embed-dings comparison. Linear discriminant analysis (LDA) is first applied to the speaker's x-vectors extracted from the training set, reducing their dimension from 512 to 150, fol-lowed by whitening, centering, and length normalization using the training x-vectors mean. Then, the speakers' x-vectors extracted from the SRE combined data (SREs 04–10) are used to train the PLDA. Furthermore, to reduce the variability between train-ing data and test data, the PLDA parameters are adapted with in-domain development data of SRE18 (Arabic Tunisian speech), by calculating the mean and covariance of the in-domain data and applying them to whiten the test x-vectors.

---

[3]The data can be downloaded from http://www.openslr.org.

9.2.4 TSP speaker verification system for the SRE'19 multimedia challenge

This section describes the experimental setup for the TSP speaker verification system for the audio part of the SRE'19 multimedia challenge. For the audio task, the speech segments from the audiovisual SRE'19 dev and test sets were extracted from the VAST corpus. For this challenge, the submitted speaker verification is a single system based on the E-TDNN architecture, which has proven its better performance compared to the baseline TDNN [15]. We start by presenting a summary of the data used to train the various components of the system, followed by a description of the back-end scoring.

For training the x-vector extractor E-TDNN 1 276 888 segments from 7 323 speakers selected from Voxceleb 1 (dev and test) and Voxceleb 2 (dev) are used. Data augmentation is performed to increase the amount and diversity of the available training data using the same method described in the CTS challenge.

For the back-end scoring, a classifier based on PLDA is used for the speaker embeddings comparison. The x-vectors extracted from the training data are centered, and their dimension is reduced to 200 components using LDA. After applying length normalization, they are used to train a PLDA model. Furthermore, the PLDA parameters are adapted using the standard Kaldi domain adaptation recipe [17] with the audio part of datasets 1, 2, and 3 (DS1, DS2, and DS3) of the BioSecure database[4] [18].

The DS1 consists of audiovisual data acquired over the Internet under unsupervised conditions by standard camera and microphone for about 600 users. DS2 set consists of audiovisual acquisition in an office environment using a desktop PC, and DS3 consists of audiovisual recording using a mobile platform in two acquisition conditions, indoor (in a quiet room) and outdoor (noisy environments). The audio part is encoded at 16 KHz, and each user is asked to pronounce a PIN code (four digits) in English and his native language, digits from 0 to 9 in English, two different phonetically rich sentences in English, and two different phonetically rich sentences in his native language.

The choice of the BioSecure database for the PLDA adaptation is based on its acquisition condition and its audio quality, which is similar to the audio part of the SRE'19 multimedia database that includes segments extracted from amateur video recordings that are downloaded from YouTube and recorded using personal devices in various acoustic backgrounds. Also, the sentences from the BioSecure database spoken in native language could contribute to solve the out-of-domain issue. We did not use the BioSecure database for the PLDA training, nor the embedding extractor training because it does not contain enough data.

---

[4]BioSecure database is available on https://biosecure.wp.imtbs-tsp.eu/how-to-get-the-database/.

Table 9.1   Data usage for training the TDNN, E-TDNN, and the PLDA for systems developed during the SRE'19 CTS and multimedia challenges

System	X-vector Training dataset	LDA/PLDA Training dataset	PLDA Adapted dataset
(0) TDNN	SRE + SWB	SRE combined	No adaptation
(1) TDNN	SRE + SWB	SRE combined	Dev-SRE18
(2) E-TDNN	SRE + SWB	SRE combined	Dev-SRE18
(3) E-TDNN multimedia	Voxceleb 1 and 2	Voxceleb 1 and 2	No adaptation
(4) E-TDNN multimedia	Voxceleb 1 and 2	Voxceleb 1 and 2	Audio-BioSecure (DS1, DS2, DS3)

9.2.5   Results for TSP speaker verification systems on the SRE'19 CTS and multimedia challenges

In this section, we present the experimental results on the SRE'19 CTS database and on the audio part of the SRE'19 audiovisual database VAST obtained using the systems described in Table 9.1. We use the scoring tool provided by the SRE'19 organizers to report the results in terms of EER, minimum detection cost function (min_C), and actual detection cost function (act_C). Usually, score normalization is applied but, in our developed systems, due to lack of time it was not implemented.

For the CTS challenge, Table 9.2 summarizes the results obtained on the SRE'19 CTS progress (dev) and evaluation (test) set. For fusion, each system provides log-likelihood scores. These scores were first calibrated and then passed to the fusion module. The calibration is achieved by applying a calibration transformation to the raw scores. The transformation parameters are learned on the dev set of CTS challenge and applied to the dev and test scores sets. Both calibration and fusion were

Table 9.2   TSP speaker verification systems performance on the progress (dev) and evaluation (test) set of SRE'19 CTS challenge for systems (0), (1), (2) described in Table 9.1 and the fusion of (1) and (2) (EER (%) | min_C | act_C)

System	Fusion	EER% dev | test	min_C dev | test	act_C dev | test
(0) TDNN		10.75	0.66	3.95
(1) TDNN		10.02 | 10.82	0.6 | 0.616	0.604 | 0.621
(2) E-TDNN		9.15 | 9.37	0.573 | 0.522	0.599 | 0.568
Fusion (1)+(2)	Linear regression	8.57 | 8.70	0.546 | 0.519	0.554 | 0.538
Fusion (1)+(2)	Pool adjacent violators (PAV) calibration	8.56 | 8.70	0.545 | 0.521	0.549 | 0.526

DET curve

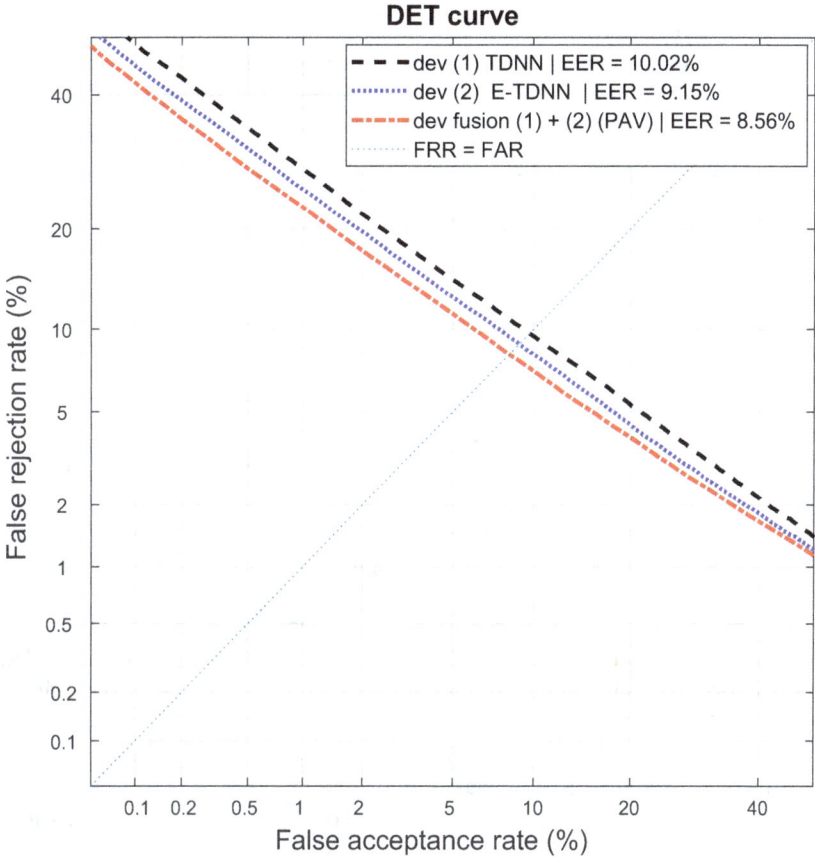

Figure 9.2　*DET performance curves on the development set of SRE'19 CTS*
database for system (1) based on TDNN, system (2) based on
E-TDNN, and the fusion of (1) and (2)

performed using Bosaris toolkit [19] based on the isotonic regression implemented
with the PAV algorithm and the standard logistic regression.

As reported in Table 9.2, systems (1) and (2) are close in performance, and as
shown in the detection error trade-off (DET) curve (Figure 9.2), the fusion of TDNN
and E-TDNN improves the performance. The EER is improved from 10.82% and
9.15% to 8.7%. Also, the analysis of obtained results shows that the main issue of
this particular challenge is the out-of-domain problem. The adaptation of the PLDA
using Tunisian Arabic speech (dev-SRE18) in systems (1) and (2) improves the
biometric performance compared to the system (0) that is based on out-of-domain
PLDA. We observe an improvement in act_C from 3.95 to 0.604 when the PLDA
back end is adapted to the target domain. The SRE'19 CTS dataset is noisy and
contains Tunisian Arabic speech. However, the databases used to train the x-vectors

Table 9.3   *TSP speaker verification systems' performance on the development*
(dev) and evaluation (test) audio part of SRE'19 multimedia challenge
for systems (3) E-TDNN (PLDA not adapted) and (4) E-TDNN (PLDA
adapted with BioSecure database) (EER (%) | min_C | act_C)

System	EER% dev | Test	min_C dev | Test	act_C dev | Test
(3) E-TDNN	14.16	0.645	0.991
(4) E-TDNN	6.95 | 7.87	0.384 | 0.409	0.408 | 0.425

and the PLDA are collected in clean conditions (SRE combined), and most of them contain recordings of English speech.

In order to reduce the variability between training and test data, the PLDA parameter adaptation is needed using in-domain data. To resolve this issue, one of the participants in this challenge collected his own Tunisian dataset with the goal of getting more in-domain speech data to train the x-vectors and the PLDA.

For the multimedia challenge, Table 9.3 presents the results of the TSP-submitted speaker verification system in the development and test set of VAST database. The utterances in the multimedia dataset contain multi-speaker audio. For the NIST SRE'19 multimedia audio dataset, the speaker labels are only provided for the enrollment data. The test audio utterances extracted from videos may contain more than one person's voice or background noise or music. The reported results in Table 9.3 are obtained without performing the diarization of test segments in dev and test set. During the evaluation, one of the participants studied the impact of diarization [20]. They reported the results obtained with x-vectors without diarization and with a diarization method [21] based on agglomerative hierarchical clustering of x-vectors. They have shown that the EER improves from 4.41% to 2.30% in the test set of VAST. The majority of submitted systems are using diarization.

In addition, comparing the results obtained for the systems (3) and (4) in Table 9.3, we observe the impact of back-end PLDA scoring in the speaker verification performance. For the system (3), where the PLDA model is not adapted to the test data, we obtain an EER of 14.6%. On the other hand, for the system (4), where the PLDA is adapted, the EER is improved to 6.95%.

For the systems (3) and (4), the architecture and the training data used to train the E-TDNN and to extract the x-vectors were the same. Both systems were based on the E-TDNN trained with the Voxceleb dataset. However, the PLDA model was not the same. The PLDA model in the system (3) was trained using x-vectors extracted from Voxceleb without adaptation of the data. On the other hand, for the system (4), the PLDA is also trained with x-vectors extracted from Voxceleb but adapted using the mean and the covariance of the x-vectors extracted from the BioSecure database that contains audio data similar to the level of acquisition condition and audio quality as the VAST database. This adaptation reduces the variability between the training (Voxceleb) and test data (VAST), and as a result improves the biometric performance from 14.6% to 6.95% in terms of EER.

9.2.6 Conclusions

Through our participation in the SRE'19 CTS challenge, we have shown that fusion of the TDNN and the E-TDNN systems yields an improvement in the biometric performance even if the systems have quite similar architecture. The EER is improved from 10.82% and 9.15% to 8.7%. Also, we observed that x-vectors have difficulties in cases of domain mismatch. For the SRE'19 CTS challenge, the main challenge was to build systems using a lot of out-domain data (mostly English) and to adapt them to the target domain using the limited available Tunisian Arabic telephone data. The improvement in biometric performance was evident when adapting the evaluation data to the target domain. During the evaluation, it was shown that x-vectors and score normalization, together with domain adaptation methods, are important to improve biometric performance results in the target domain with very limited domain data.

For the multimedia challenge on the VAST database, we focused our work for the adaptation of the PLDA rather than fusion of multiple systems. We showed the impact of the adaptation of the PLDA model trained with x-vectors extracted from Voxceleb with the audio part of the BioSecure database. As far as we know, this is the first time where the BioSecure database was used for adaptation purposes. This database is one of the few databases that has recordings 16 kHz, collected with different acquisition conditions (calm, noise, mobile, microphone) and variability of voice information, which make it adapted to solve the issue of mismatched domain data. Also, during our participation, we have confirmed that when dealing with multi-speaker test utterances, the diarization is mandatory to obtain better results.

Moreover, the new challenge for multimedia was the audiovisual speaker verification on VAST data. The following sections describe the TSP face recognition system developed for the fusion with the speaker verification system.

9.3 TSP face recognition system for SRE'19

In this section, we give a survey of some prominent face recognition systems. We then provide a description of our face verification system submitted to the NIST SRE'19 multimedia evaluation while presenting the key modifications introduced to achieve better performance in the challenge.

9.3.1 Survey of face recognition systems

Currently, the state-of-the-art facial biometric algorithms are based on deep convolutional neural networks (CNNs). Table 9.4 summarizes the most prominent published DNN-based facial recognition systems. Most of them are either proprietary, only a description of the system is provided, or trained on private databases. To compare biometric systems objectively, it is mandatory to use the same database and

Table 9.4   Summary of the state-of-the-art DNN-based face recognition systems

System	Size of training database (millions of images)	Accuracy on LFW ± Std (%)	Reproducibility
Camvi [24]	5.00	99.87± 0.18	No
Ever.ai [24]	10.00	99.85± 0.20	No
FaceNet [26]	260.00	99.63± 0.09	No
DeepID2 [28]	0.29	99.52± 0.12	No
VGG-DeepFace [29]	2.60	98.95*	Yes
DeepFace [27]	4.40	97.35± 0.25	No
CasiaNet [30]	0.50	96.13± 0.30	No
OpenFace [31]	0.60	92.92± 1.34	Yes

*Std is not reported by original authors.

the same testing protocols [22]. To this end, we will report the performance of the systems on the Labeled faces in Wild database (LFW) [23], where the comparison metric is the pair matching accuracy using the 10-fold cross-validation protocol.

Some best performing systems in the face recognition ecosystem are described in the next paragraphs.

The **Camvi's** model [24] is trained on a subset of MS-celeb-1M face database, containing around 80 k identities and 5 million (M) faces. The authors tried to remove the overlapped faces in LFW with close similarity scores. The recognition model is a single CNN with a size of 230 MB, which outputs an embedding vector with 256 float point numbers for an input image. L2 distance is used to measure the similarity between two feature vectors and compute average accuracy for each subset using the best threshold from the rest of the nine subsets of the ten folds.

Ever.ai [24] trained their model on a private photo database with no intersection with LFW. They also trained custom face and landmark detectors for preprocessing and built their primary face recognition model on a database containing over 100 k identities and 10 M images. The recognition model is a single deep ResNet model [25], which outputs an embedding vector given an input image, and the similarity between a pair of images is evaluated via an L2-norm distance between their respective embeddings. The system is a proprietary system with no extensive description provided by the authors. We report its performance on the LFW benchmark because it is one of the best performing systems.

FaceNet [26] was developed by Google. It is a unified system for face verification, identification, and clustering. It extracts Euclidean representations from images with the advantage of being general purpose. The features are also compact (with a dimension of 128) compared to traditional representations (Gabor features, for example). The system was trained on a huge private database of 260 M images from 8 M subjects. It was trained for 1 000 hours.

DeepFace [27] is developed by Facebook. It processes images in two steps. First, it corrects the angle of a face so that the person in the picture becomes forward-facing, using a three-dimensional model of an "average" forward-looking face. The second step is to propagate the face to the DNN in order to extract its representation. The system was trained on a private database consisting of 4.4 M images from 4 k subjects (average of 1 k per subject). It has 97.35% accuracy on LFW.

DeepID2 [28] was developed by the Department of Information Engineering of the Chinese University of Hong Kong. The features are learned using deep convolutional networks. The face identification task increases the interpersonal variations by drawing apart DeepID2 features extracted from different identities. In contrast, the face verification task reduces the intrapersonal variations by pulling DeepID2 together extracted from the same identity, both of which are essential to face recognition. It was trained on a private database consisting of 200 k images from 10 k subjects. Compared to other databases such as Google's or Facebook's systems, the size of the database can be considered relatively small. It gives 99.15% verification accuracy on the LFW database.

VGG-DeepFace [29] was developed by the Visual Geometry Group (VGG) from the University of Oxford. The system was trained on 2.6 M images containing 2.6 k identities. The published performance on LFW is 98.95%. The VGG system is essentially a very deep CNN. It leverages two distinct methods for the training: N-way classification and triplet embedding. In the case of this system, the N-way has the advantage of faster training, while, on the other hand, triplet embedding gives a better overall performance.

CASIANet [30] was developed by the Institute of Automation, Chinese Academy of Sciences (CASIA). The system is inspired by many new successful networks, including very deep architecture, low-dimensional representation, and multiple loss functions. It was trained on the publicly available CASIA database (500 k images representing 10 k identities). The reported performance of the system on LFW is 96.13%.

OpenFace [31] is an implementation of the FaceNet system based on [26]. The source code is publicly available as well as the trained model. It was trained on the publicly available CASIA-webfaces and FaceScrub databases. The system has 92.92% accuracy on LFW.

Table 9.4 provides a summary of the performance of some of the face recognition systems studied. The results are reported on the LFW benchmark. The benchmark comprises 6 000 tests, divided into 10 partitions (folds). The performance is reported in terms of average accuracy and the standard deviation over the ten folds. When computing the accuracy for each fold, one must use the remaining nine folds in order to determine the threshold that gives the highest accuracy on the nine folds. Afterward, the threshold is applied to the remaining fold giving the accuracy on that folds. This process is repeated for each fold separately. The accuracies are then averaged to provide the reported metric along with the standard deviation. Further details on the LFW database are provided in Section 9.3.3.

All the mentioned methods use DNNs. The reproducibility of these systems hangs mainly on the availability of the training data. For example, even when the architecture is described in detail as in the case of FaceNet, OpenFace, which tried to reproduce FaceNet's performance, achieved only 92.92% in comparison with the 99.6% accuracy of FaceNet. This might be due to FaceNet being trained on 260 M images in contrast to the 600 k of OpenFace. We suspect that another reason for the difference in performance relates to the quality of the face detector. Our face recognition system, described in the following section, is built upon the OpenFace framework. The main reason behind the choice of OpenFace is reproducibility.

9.3.2   TSP face recognition system pipeline

In this section, we describe the face recognition system used in the submission for the multimedia challenge during SRE'19. Figure 9.3 shows the block diagram of the face recognition system, which is built using open-source implementations of: (1) a face detector named RetinaFace [32] and (2) a face-embedding extractor based on FaceNet [26] (built using the OpenFace implementation [31]).

We trained the face detector using the WIDER FACE dataset with the default configuration described in [32]. As for the face-embedding extractor, we used the MS-celeb-1M dataset for training. After removing label noise using the density-based spatial clustering of applications with noise (DBSCAN) clustering algorithm, the training dataset comprises 80 000 subjects with a total of 4 000 000 images.

For the enrollment, we select the frames provided by the manual annotations given by NIST. We then crop the frame to the bounding box specified in the meta-data. The face is then detected inside the provided bounding box. DLIB [33] land-mark detector is used to obtain the face landmarks. We use the outer eyepoints and the nose tip in order to align the face to a predefined layout. The aligned image is resized to 96 × 96 pixel rectangle and fed to the face-embedding extractor. The result is a 512 component face embedding per frame.

As for the test videos, we begin the processing by extracting one frame per 0.5 seconds from each video. Then we apply the RetinaFace face detector to each extracted frame to get all the faces present in the frame. We then run the land-mark detector on each bounding box found by the face detector. After aligning and

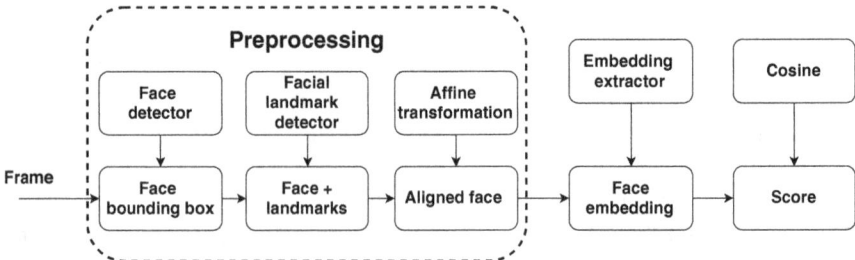

Figure 9.3   *Block diagram of the TSP face recognition system for SRE'19 multimedia challenge*

resizing the faces, we extract the embedding of each face. In order to compute a single score for each trial involving an enrollment video and a test video, we compute the maximum of the cosine similarity scores obtained by comparing all of the enrollment embeddings and the embeddings of the faces found in the video. The scores are not postprocessed.

In the following subsections, we present the training and validation databases used in the face recognition system. We also detail the system's components and explain the improvement provided by each modification to the OpenFace framework.

9.3.3 Databases used in the TSP face recognition system

To train, test, and validate our face recognition system, we used multiple databases. The databases that we used are WIDER FACE for training the face detector; MS-celeb-1M for training the embedding extractor; and LFW and AgeDB for validation.

The LFW dataset contains 13 233 target face images with a considerable degree of variability in facial expressions, age, race, occlusion, and illumination conditions; 1 680 of the people pictured have two or more distinct photos in the dataset. The only constraint on these faces is that they were detected by the Viola-Jones face detector [34]. The protocol specifies two views of the dataset. View 1 is for model selection and algorithm development. It contains two sets: 1 100 pairs per each class (matched/mismatched) for training and 500 pairs per each class for testing. View 2 is designed for performance reporting. It is divided into 10 sets (folders), each with 300 matched pairs and 300 mismatched pairs. The cross-validation evaluation can be adopted among these ten folders. The final verification performance is reported as the mean recognition rate and standard error over the ten folds. It has to be noted that the task is to do pair matching. Given a pair of images, the goal is to decide whether they belong to the same subject. This task is similar to face verification, except that the evaluation metrics proposed by the database collectors is the accuracy of the pair matching.

WIDER FACE [35] dataset is a face detection benchmark dataset, of which images are selected from the publicly available WIDER dataset; 32 203 images are chosen with 393 703 labeled faces having a high degree of variability in scale, pose, and occlusion. WIDER FACE dataset is organized based on 61 event classes (i.e., parade, meetings, protests).

AgeDB [36] is a manually collected, in-the-wild age database containing images annotated with accurate-to-the-year, noise-free labels. As demonstrated by a series of experiments utilizing the state-of-the-art algorithms, this unique property renders AgeDB suitable when performing experiments on age-invariant face verification, age estimation, and face age progression "in-the-wild". The database contains 16 488 images from 568 subjects.

The MS-celeb-1M is one of the largest publicly available databases. It has 100 k subjects and almost 10 M images. Popular search engines are used to provide about 100 images for each subject. The images are collected based on their metadata, not their content. This results in the dataset having a considerable amount of noise. The dataset is constructed by Microsoft and is available for noncommercial use. Guo *et al.* [37] further describe the process of assembling the images and the metrics used for the choice of the 100 k celebrity provided in the dataset. We used the whole dataset for training the neural network. The MS-celeb-1M database contains a significant portion of mislabeling because it was collected automatically using web crawlers. In order to improve the performance, we leveraged clustering algorithms to clean the database. We applied DBSCAN [38] to reduce the mislabeling of the database. We worked under the assumption that there is no overlap between the identities of the labels provided in the database metadata. In other words, under the same label we can find multiple identities, but there is no overlap between the identities belonging to different labels. As the number of the identities in each label is unknown, we proceed by applying DBSCAN clustering algorithm onto each label. The clustering is done on the embeddings computed using our model from [39]. The cluster with the highest number of samples is kept and the remaining clusters are discarded. In cases where the number of samples in the biggest cluster is lower than three, the label is discarded. Furthermore, the MS-celeb-1M database has a bias toward the LFW dataset as there is an overlap of the identities between both databases. To reduce this bias, we removed the labels (identities) that have an Euclidean distance lower than 1.2 from any sample from the LFW. Note that the threshold with which the accuracy is computed on the LFW benchmark in [39] is around one. We used a stricter threshold because the model from [39] is trained on a non-cleaned version of MS-celeb-1M and presents a bias to LFW. Thus, the cleaning resulted in reducing the training database to 80 000 identities from the 100 000 users provided in the MS-celeb-1M and reducing the total number of images from 10 M to 4.5 M.

9.3.4   Face preprocessing

Before using the DNN to construct the face embedding, the image containing the face should be preprocessed. The preprocessing consists of a geometric alignment of the face. The alignment contains two steps. The first step is to detect the bounding box of the face. Once the face is detected, we need to detect the facial landmarks, which in our case are the 68 facial points defined by the Multi-PIE 68 points markup shown in Figure 9.4. The landmarks that are used for normalization are the eyes and the nose. Using these landmarks, the face is rotated, scaled, and cropped. The resulting image has 96×96 pixels. Figure 9.5 shows the effects of preprocessing on one image.

The alignment used in the preprocessing of the training set should be applied in the enrollment and verification phase in the same manner. To achieve the best performance, the same face detector and landmark detector used in preprocessing the training data should be used when exploiting the DNN.

Figures 9.6 and 9.7 show examples of the alignment using the affine transformation on samples of good as well as bad quality. When the subject's face is not facing forward and presents a high degree of rotation, the alignment results in a stretched face.

The first challenge in the face alignment phase is face detection. This step constitutes a high impact on the overall performance of the face recognition system.

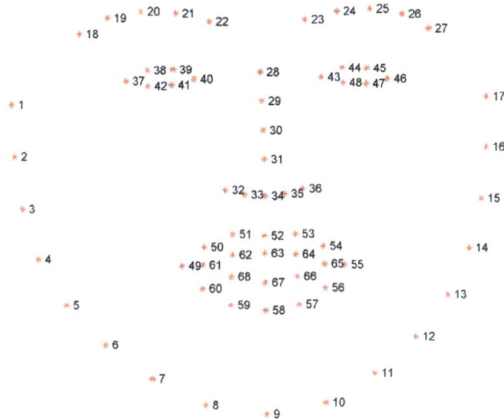

Figure 9.4   The Multi-PIE 68 points markup [40] used for face landmark annotation

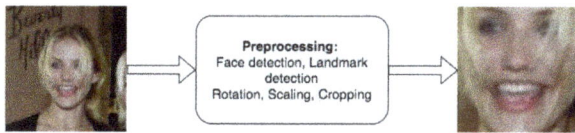

Figure 9.5   Example of the preprocessing of an image from LFW using eyes and nose positions

This impact is especially visible when the acquisition conditions are adverse (shadows, partial face, occlusion).

To study this impact, we tried different face detection methods ranging from classical methods such as the Viola–Jones [34] algorithm to newer ones based on DNNs such as Single Shot MultiBox Detector (SSD) [41] and RetinaFace [32]. As shown in Table 9.5, the choice of the face detector has a noticeable impact on the performance of the whole system.

In order to apply the affine transformation using the position of the outer points of eyes and nose, we need to detect the facial landmarks. In order to understand the impact

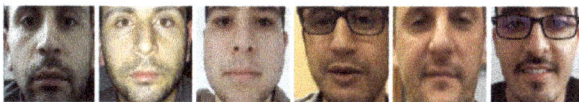

Figure 9.6   Examples from the alignment of images from the ATSIP'2018[1] database with good acquisition conditions, i.e., frontal face, good illumination

[1] ATSIP'2018 is a database acquired in the H2020 SpeechXRays project by TSP.

Figure 9.7   Examples from the alignment of images from the ATSIP '2018[5] database with bad acquisition conditions, i.e., face turned to a great degree

Table 9.5   Biometric performance of the face recognition system using different face detectors. Face landmark detection is done using DLIB implementation of ERT. Face embeddings are extracted using the FaceNet architecture trained on the cleaned version of MS-celeb-1M

Face detection method	Accuracy on LFW (%)	EER on SRE'19 multimedia dev (%)
Viola–Jones	97.53	17.00
SSD	98.82	14.36
RetinaFace	99.32	11.20

Table 9.6   Biometric recognition performance of the face recognition system using different face landmark detection methods. Face detection is done using the SSD model. Face embeddings are extracted using the FaceNet architecture trained on the cleaned version of MS-celeb-1M

Landmark detection method	Accuracy on LFW (%)	EER on SRE'19 multimedia dev (%)
ERT DLIB implementation [42]	98.82	14.36
2DFAN [43]	98.90	14.20
CNN (trained on ibug 300W)	98.68	14.56

of the quality of the landmark detector, we used three landmark detectors: ensemble of regression trees (ERT) proposed by [42] and implemented in the DLIB [33] toolbox; a 2DFAN DNN-based solution introduced by [43]; and a CNN that we trained on ibug 300W [44]. From Table 9.6, we can conclude that the impact of the face landmark detector is negligible on the overall performance of the face recognition system. This might be due to the embedding extractor neural network being trained using a training set with the same face landmark detector. In fact, the DNN gets used to the errors induced by the landmark detector. What is important is to use the same landmark detector used for training when exploiting the system for face verification.

9.3.5   Embedding extractor

The embedding extractor used in the face recognition system is a deep convolution neural network. The DNN architecture used in OpenFace is an implementation of the FaceNet model based on [26]. It was inspired from the inception network [45].

Initial version of the DNN architecture

The initial architecture that we used consists of an input layer, an output layer, and 24 hidden layers among which there are seven inception layers. The initial version of the network counts 3 733 968 parameters. The DNN extracts feature vectors that give the best possible separation between subjects. It uses triplet embedding to optimize the representations. Schroff *et al.* [26] detail the process of triplet selection and optimization. The loss function defined in (9.5) is based on the triplet loss optimization scheme that consists of choosing two samples from the same class (the anchor and the positive) and a sample for a different class (the negative). The goal of the training is to separate the anchor-positive pairs from the anchor-negative pairs by at least the margin α. The triplet mining is done online by selecting the triplets that do not follow the rule given by (9.6). We select the hard negative triplets where the anchor-negative distance is less than the anchor-positive distance.

$$L(\theta) = \sum_i^N \left[\left\| f(x_i^a) - f(x_i^p) \right\|_2^2 - \left\| f(x_i^a) - f(x_i^n) \right\|_2^2 + \alpha \right] \qquad (9.5)$$

In (9.5), θ represents the network parameters, x_i^a is the anchor sample, x_i^p the positive sample, and x_i^n the negative sample for subject i. $f(x)$ is the DNN representation of the image x. In order for the training to be efficient (to save computing time), only the triplets that verify Equation 9.6 rule are selected, as other triplets will not improve the network performance.

$$\left\| f(x_i^a) - f(x_i^p) \right\|_2^2 - \left\| f(x_i^a) - f(x_i^n) \right\|_2^2 + \alpha > 0 \qquad (9.6)$$

This selection process allows for the training to run faster and be more efficient because we will not need to back-propagate triplets that have little effect. If a triplet does not verify the inequality from (9.6), then the considered samples contain too little intra-class variance and a high inter-class variance. As a result of such training, the network outputs a low-dimensional representation of an input image, which consists of a normalized feature vector of size 128. This representation can be leveraged to do either verification, identification, or clustering. The SRE'19 multimedia challenge focuses on verification performance.

Final version of the DNN architecture

The final version used in the SRE'19 multimedia submission follows the same approach as the initial DNN model. However, we introduced the following modifications:

- The inception-v3 layers were replaced by inception-ResNet-v1 [46] layers. Inception-ResNet-v1 has a computational cost that is similar to that of inception-v3 and allows for faster training.
- The triplet loss function was modified according to (9.7). The modified version takes into account the cosine similarity between the embeddings. To compensate for the added terms, α is changed from 0.2 to 0.6. The value 0.6 was chosen empirically.
- The size of the embeddings was changed from 128 real components to 512 by trial. The choice of the embedding size was carried out using the LFW benchmark. We chose the size that gave the best accuracy on LFW.

$$
\begin{aligned}
L(\theta) &= \sum_i^N [\|f(x_i^a) - f(x_i^p)\|_2^2 - \|f(x_i^a) - f(x_i^n)\|_2^2 - f(x_i^a) \cdot f(x_i^p) \\
&+ f(x_i^a) \cdot f(x_i^n) + f(x_i^p) \cdot f(x_i^n) + \alpha]
\end{aligned}
\tag{9.7}
$$

where "$\cdot$" is the dot product. The Euclidean distance component and the cosine distance have an overlap. We do not remove the overlapping components of the loss function in order to give more importance to the distance between the anchors and the negatives samples than the distance between the positives and the negatives sampled. If we develop (9.7), we obtain:

$$
\begin{aligned}
L(\theta) &= \sum_i^N [\|f(x_i^n)\|_2^2 + f(x_i^n) \cdot f(x_i^p) + 3f(x_i^a) \\
&\cdot f(x_i^n) - 3f(x_i^a) \cdot f(x_i^p) - \|f(x_i^p)\|_2^2 + \alpha
\end{aligned}
\tag{9.8}
$$

The impact of the improvements to the DNN is not evident on the LFW dataset; however, on the AgeDB-30 benchmark and on the SRE'19 multimedia development partition, the improvement in the performance is much more visible. On the AgeDB-30 benchmark, the accuracy improved from 89% to 97% as shown in Table 9.7. This shows that the improved architecture is more robust to age variance than the initial version provided by the OpenFace framework.

Figure 9.8 reports the performance of the submitted face recognition system on the dev and test partitions of the multimedia challenge. The curves show that the submitted system performs better on the test set than on the dev set. This might be due to the higher number of tests in the test partition in comparison with the dev partition. In fact, the number of tests in the test partition is 12 times higher. The

Table 9.7   Biometric recognition performance of the studied DNN architectures. The face detection is done using the RetinaFace face detector. Face landmark detection is carried out using the DLIB implementation of ERT. The DNN is trained on the cleaned version of MS-celeb-1M

	Accuracy on LFW (%)	**Accuracy on AgeDB-30 (%)**	**EER on SRE'19 multimedia dev (%)**
Initial version	99.32	89.00	11.20
Final version	99.80	97.00	4.36

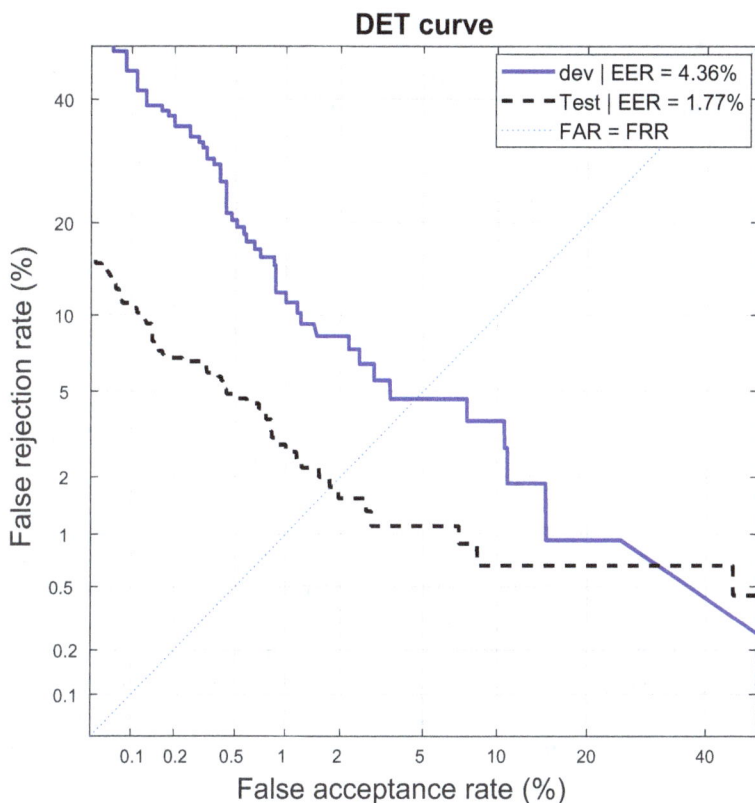

Figure 9.8   *DET curve performance of the submitted face recognition system on the dev and test partitions of the multimedia challenge*

difference in the number of tests also results in higher error margin for the dev set. For example, the error margin at the EER is 1.89% for dev set and 0.55% for the test set.

The scores provided by our face recognition system are computed using the cosine distance. These scores are suitable for the EER metric. However, they cannot be used to compute the actual cost (act_C), which is the primary metric for the SRE'19 challenge. The act_C metric is based on the log-likelihood ratio (LLR) scores. The scores were not postprocessed (no Z-norm, T-norm, etc.) but were calibrated using the Bosaris toolkit [19]. The score calibration is done in the same step as the fusion. The multimedia challenge is an audiovisual task. The face recognition scores need to be fused with the speaker recognition scores.

9.3.6   Conclusions

In this section, we presented the TSP face recognition system submitted in the SRE'19 multimedia evaluation. The system is built upon the OpenFace framework, to which we introduced several modifications to obtain better performance on the development set of the VAST corpus.

Among the modifications applied to the framework, the use of the RetinaFace face detector resulted in the most significant improvement in performance on the development set. The quality of the detected face landmarks depends significantly on the correctness of the bounding box given by the face detector. Using the correct face landmark results in better face alignment and more robust templates. It is worth noting that this detector, RetinaFace, was also used by other submissions, which shows its success in the task of face detection.

Although the majority of the submissions to the challenge used CNN models for face recognition, a system submitted by one team used TDNN for face recognition, which gave 19.84% EER on the development set. This shows that CNN is one of the better-suited architectures for face recognition.

We also note that applying whitening and using PLDA for scoring does not improve the recognition performance, at least in our submission.

Finally, applying enrollment filtering using some quality measures is detrimental to the performance of the face recognition system. If the enrollment reference is of bad quality, the comparison against good test references will result in lower similarity scores and, as a result, impacts the decision threshold. Some participants implemented enrollment provisioning procedures, where they locate the target face in frames other than the keyframes given by the metadata to obtain a better enrollment reference. However, in our submission, we only used the keyframes with the best quality of the enrollment videos. This helped us obtain the best performance in the EER metric between the 14 submitted systems to the SRE'19 multimedia challenge.

In the next section, we explain how the output of the face recognition system and the speaker recognition system is fused and how the system performs compared to the rest of the submissions.

9.4   Audiovisual biometric system for the SRE'19 multimedia challenge

In this section, we present the TSP audiovisual system developed during the SRE'19 multimedia challenge. Figure 9.9 shows the block diagram of this system. It is based on the fusion of single speaker verification system based on the E-TDNN described in Section 9.2.4 and face recognition system described in Section 9.3. Table 9.8 reports the results obtained on the dev and test set of the multimedia database.

For the audio results, compared to the submitted systems in the challenge, we obtained 6.95% EER on the development set with only a single system due to the use of the BioSecure database for the adaptation of the PLDA model, knowing that the

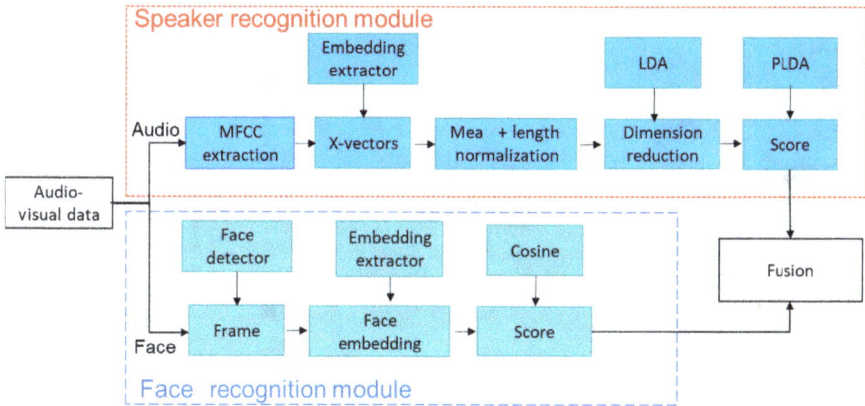

Figure 9.9　*Block diagram of TSP's audiovisual system for NIST 2019 multimedia challenge*

best EER of the challenge was 4.69% on the dev set. However, the results in terms of min_C and act_C are worse than other submitted systems and this is expected because we have not applied the diarization for test utterances and the normalization methods for the scores.

For the face results, we obtain the best performance on the VAST corpus when using the EER metric. As shown in Table 9.8, we obtain 4.36% EER on the development set whereas the second best submission gives 6% EER. On the evaluation set, we get 1.77% EER, which is also the best EER among the submitted systems. When the results are interpreted in terms of the primary metric, act_C, our face system becomes the fifth best performing system among each team's best submission as well as the primary submissions (see Figure 9.10, our system is denoted as T14). The difference in performance between the EER and act_C can be attributed to score calibration and score normalization. In fact, score calibration does not impact the EER of the system. It only impacts the cost of the detection.

During the challenge, we have submitted two systems named primary and secondary based on two strategies for the fusion. For the primary system, the fusion is performed using isotonic regression implemented with the PAV algorithm. Scores were

Table 9.8　*TSP audiovisual systems results on the dev and test set of SRE'19 multimedia database (EER (%) | min_C | act_C)*

System	Approach	EER% dev	Test	min_C dev	Test	act_C dev	Test
Speaker	x-Vector	6.95 | 7.87	0.384 | 0.409	0.408 | 0.425
Face	FaceNet	4.36 | 1.77	0.286 | 0.101	0.295 | 0.103
AV-Primary	PAV calibration	0.93 | 1.99	0.090 | 0.14	0.090 | 0.149
AV-Secondary	Linear regression	4.41 | 1.77	0.188 | 0.059	0.201 | 0.071

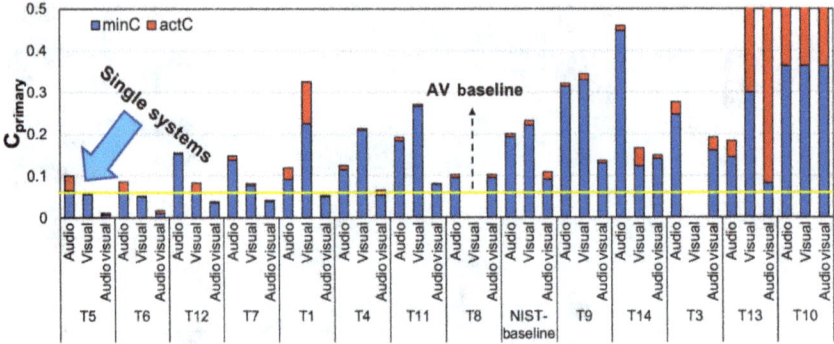

Figure 9.10   *Performance of the primary submissions for all three tracks (i.e., audio, visual, and audiovisual tracks) of the audiovisual SRE'19 [1], in terms of the minimum (in blue) and actual (in red) detection costs. The top performing audio and visual systems are both single systems (i.e., no fusion)*

first calibrated, fusion performed, and then a final calibration step performed. For the secondary system, the fusion is performed using standard logistic regression. Scores sets were first calibrated, and then fusion is performed. The calibration parameters for both systems were trained using the SRE'19 multimedia dev set for application to the evaluation scores. Bosaris toolkit [19] was used for fusion and calibration.

The fusion of the scores in the primary submission is done in a nonparametric way. We independently choose the value for each point, subject only to the monotonicity constraint. This is done using an isotonic regression by the PAV algorithm. On training data, this fusion is optimal, no matter how optimality is measured. The type of score distribution is not important. The procedure is invariant to any monotonic warping of the scores. In contrast, the parametric logistic regression calibration solution works best for approximately normal score distributions [19].

The logistic regression used in the secondary submission is done according to (9.9):

$$S_{\text{fused}} = a + \sum_{i}^{N} w_i S_i \tag{9.9}$$

where

- S_{fused} is the fused and calibrated output.
- a is an offset.
- N is the number of subsystems.
- w_i is the weight.
- S_i is the score for each subsystem.

DET curve

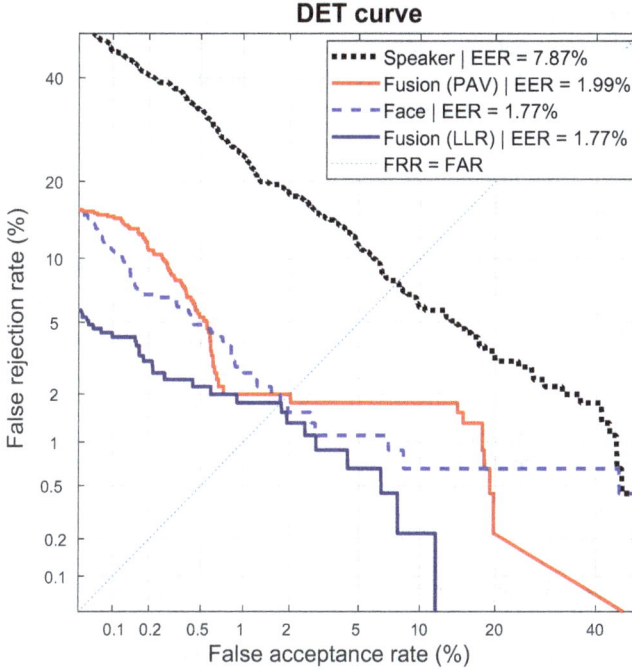

Figure 9.11   DET curve performance of the primary (with PAV) and secondary (with LLR) fusion of the TSP's submission systems on the test partition of the multimedia challenge

The parameters optimized using logistic regression are the offset a and the weights w_i. The Bosaris toolkit uses a general-purpose, unconstrained convex optimization algorithm to train the logistic regression fusion and calibration solutions. It uses the trust region Newton conjugate gradient algorithm for large-scale unconstrained minimization [47].

The results obtained for the audiovisual speaker verification on the multimedia data show an improvement of the recognition performance of the overall system (see Figure 9.11), in particular for the primary metric (act_C). The primary submission that used the PAV calibration falls in over-fitting on the development set. The secondary system, for which we did a regular fusion using linear regression, shows better performance on the test set even if it was worse on the development set.

The difference in the performance for both systems can be attributed to the difference in the size of the sets. In fact, the development contains only 52 subjects and 5 618 tests, which results in lower reliability of the results. On the other hand, the test set contains 149 subjects with 67 348 tests. It seems that we over-tuned our primary submission on the development set, resulting in degraded performance, with less robust system that could not generalize correctly.

Our fusion strategies rely on fusion of individual biometric systems scores. Evidence from multiple sources can also be combined on different levels, such as sensor, feature, or decision level. Besides, promising strategies are presented in the literature [48], investigating intermediate-level fusion, when uni-biometric features are jointly trained to exploit the correlation of audio and face by presenting both modalities with a single audiovisual representation.

9.5 Conclusions and perspectives

In the first part of this chapter, we presented the TSP speaker recognition system submitted to the NIST SRE'19 CTS challenge evaluated on the CMN2 corpus containing 8 kHz CTS in Tunisian Arabic. The purpose of this challenge was to verify the person from their voice.

The TSP audio system consisted of a score fusion of the TDNN and E-TDNN systems-based x-vector extractor followed by a PLDA classifier. The results show the importance of adapting the PLDA model to the target test data, which was the main challenge during the CTS challenge. Most of the available databases used to train the audio system containing recordings of English speech collected in clean conditions (SREs 04–10) or in the microphone channel (Voxceleb). However, the CTS database is noisy and contains Tunisian Arabic speech, and the data was limited. To resolve this issue, the team with the best performance during this challenge has collected its Tunisian database to get more in-domain speech data. Moreover, research on domain adaptation was presented during the SRE'19, such as the adversarial adaptation based on generative adversarial networks, which aims to discriminate between source and target domains (English and Arabic).

Also, in this chapter, we presented the TSP face and speaker recognition system submitted to the NIST' 2019 multimedia challenge. The goal of this challenge was the recognition of a person in videos via their voice and their face. For this, the audio and visual modalities are processed using DNN systems, and then score fusion is performed to provide an authentication decision.

For the audio part of the audiovisual submission, the system consisted of single E-TDNN, followed by a PLDA classifier adapted with BioSecure database to the multimedia audio data. The results obtained on the multimedia audio set showed that the critical point for improving biometric performance is not the fusion of several audio systems, but the key is how to solve the domain mismatch problem. In our audio system, the right choice of the database (BioSecure), used for adapting the PLDA to reduce the variability between training data and multimedia test data, has shown an impact on the improvement of biometric performance even if the speaker recognition system is based on a single E-TDNN.

For face recognition, we used an inception-ResNet-based network optimized with the triplet loss function. Same as the speaker recognition system, we use a single face recognition system in our submission. In this system, we leverage a state-of-the-art face detector, RetinaFace, which improves the alignment of the faces in the preprocessing step thus improving the overall performance of the system. Our system gives state-of-the-art performance on multiple benchmarks such as LFW and AgeDB-30. The system generalizes well on the multimedia corpus where we achieve the best performance in terms of EER on both the development set and the evaluation set. What distinguishes our submission from other submissions is that we filtered the enrollment images. The enrollment images are filtered based on the quality of the face. Partial faces, faces obstructed by an object such as sun glasses, or faces with illumination problems were removed from the enrollment samples based on the face detector confidence. However, when comparing our system to the rest of the submissions in terms of the primary metric, act_C, it achieves the fifth best results. This is due to the calibration step in which we only applied logistic regression without score normalization. Score normalization showed improvements in act_C in the submissions that applied it.

Finally, the audiovisual results indicate that the combination of voice and face recognition systems improves speaker recognition system's performance. Our face recognition system outperforms the speaker recognition system. One reason could be that the state of the art in face recognition is much more advanced than speaker recognition. In fact, faces are mainly represented as still images with no temporal content, which makes it easy to adapt the image recognition techniques where research is more extensive. The complexity of the speaker task (diarization, out of domain data, noise, etc.) is another reason for the difference in performance between speaker recognition and face recognition systems. Nevertheless, the fusion improved the results by halving the primary metric. However, there was no improvement in the EER. This shows that the impact of the fusion between the face and the voice, in our case, is mainly due to the quality of the scores and the cost of the detection, not on the discriminative power of the system. The calibration provided by the logistic regression optimizes the score distribution for a particular P_{target} and does not impact the DET curve. In addition to the improvements in the detection cost, the fusion of both modalities adds robustness to the system. One of the major weaknesses of face recognition system is that they are easily spoofed. By using the voice, we add a time dimension for the data, which improves the system resistance to presentation attacks. Also, speech can help implementing anti-spoofing measures such as prompted speech or replay detection.

In conclusion, the combination of voice and facial biometrics brings improvements in the biometric performance, providing the possibility of implementing different anti-spoofing strategies (voice, face, or synchrony detection). In such a way the audiovisual biometric systems also provide better security. Moreover, for the audiovisual biometric recognition system, one of the important points is the selection of the level at which the fusion is done. Fusion strategies generally refer to the sensor, feature, score, or decision level. Currently, with recent revolution driven by deep-learned representations for audio and face, new strategies of audiovisual biometric fusion are presented [48]. The idea is to exploit the correlation between voice and face modalities to perform a cross-modal training where each uni-biometric model is supported by the biometric model of the other modality in order to improve the effectiveness of its feature representations. Then, the modalities are combined by merging their feature representations into a single embedding vector instead of using them individually. Applying such new fusion strategies could be an interesting subject of research for future work.

Acknowledgments

This work is partially supported by the SpeechXRays and Empathic projects that have received funding from the European Union's Horizon 2020 research and innovation program, under Grant Agreements Nos. 653586 and 769872.

The authors would like to thank the anonymous reviewer for his/her careful reading and valuable and insightful comments and suggestions.

References

[1] Sadjadi S.O., Greenberg C., Singer E., *et al.* 'The 2019 NIST audio-visual speaker recognition evaluation'. Proc Speaker Odyssey; Tokyo, Japan; 2020.

[2] Tracey J., Strassel S. 'VAST: A corpus of video annotation for speech technologies'. LREC 2018 – 11th International Conference on Language Resources and Evaluation; 2019. pp. 4318–21.

[3] Brummer N. *Measuring, Refining and Calibrating Speaker and Language Information Extracted from Speech.* Stellenbosch: University of Stellenbosch; 2010.

[4] Reynolds D.A., Quatieri T.F., Dunn R.B. 'Speaker verification using adapted Gaussian mixture models'. *Digital Signal Processing.* 2000;10(1–3):19–41.

[5] Kenny P., Boulianne G., Ouellet P., Dumouchel P., *et al.* 'Joint factor analysis versus eigenchannels in SPEAKER recognition'. *IEEE Transactions on Audio, Speech and Language Processing.* 2007;15(4):1435–47.

[6] Dehak N., Kenny P.J., Dehak R., Dumouchel P., Ouellet P., *et al.* 'Front-end factor analysis for SPEAKER verification'. *IEEE Transactions on Audio, Speech, and Language Processing.* 2010;19(4):788–98.

[7] Richardson F., Reynolds D., Dehak N. 'Deep neural network approaches to SPEAKER and language recognition'. *IEEE Signal Processing Letters.* 2015;22(10):1671–5.

[8] Snyder D., Ghahremani P., Povey D., *et al.* 'Deep neural network-based speaker embeddings for end-to-end speaker verification'. 2016 IEEE Spoken Language Technology Workshop (SLT). IEEE; 2016. pp. 165–70.

[9] McLaren M., Lei Y., Ferrer L. 'Advances in deep neural network approaches to speaker recognition'. 2015 IEEE international conference on acoustics, speech and signal processing (ICASSP). IEEE; 2015. pp. 4814–18.

[10] Ghalehjegh S.H., Rose R.C. 'Deep bottleneck features for i-vector based text-independent speaker verification'. 2015 IEEE Workshop on Automatic Speech Recognition and Understanding (ASRU). IEEE; 2015. pp. 555–60.

[11] Snyder D., Garcia-Romero D., Sell G., *et al.* 'X-vectors: robust DNN embeddings for speaker recognition'. 2018 IEEE International Conference on Acoustics, Speech and Signal Processing (ICASSP); 2018. pp. 5329–33.

[12] Villalba J., Chen N., Snyder D., *et al.* 'State-of-the-art SPEAKER recognition with neural network embeddings in NIST SRE18 and speakers in the wild evaluations'. *Computer Speech & Language.* 2020;60(4):101026.

[13] Snyder D., Garcia-Romero D., Sell G., *et al.* 'Speaker recognition for multi-speaker conversations using x-vectors'. ICASSP 2019-2019 IEEE International Conference on Acoustics, Speech and Signal Processing (ICASSP). IEEE; 2019. pp. 5796–800.

[14] Povey D., Ghoshal A., Boulianne G. 'The Kaldi speech recognition toolkit' in IEEE 2011 Workshop on Automatic Speech Recognition and Understanding'. *CONF. IEEE Signal Processing Society*; 2011.

[15] Villalba J., Chen N., Snyder D., *et al.* 'State-of-the-art speaker recognition for telephone and video speech: the JHU-MIT submission for NIST SRE18'. Interspeech; 2019. pp. 1488–92.

[16] Snyder D., Chen G., Povey D. 'Musan: a music, speech, and noise corpus'. *arXiv preprint arXiv:151008484*. 2015.

[17] Bousquet P.M., Rouvier M. 'On robustness of unsupervised domain adaptation for speaker recognition'. Interspeech; 2019. pp. 2958–62.

[18] Ortega-Garcia J., Fierrez J., Alonso-Fernandez F., *et al.* 'The multiscenario multienvironment biosecure multimodal database (BMDB)'. *IEEE Transactions on Pattern Analysis and Machine Intelligence*. 2010;32(6):1097–111.

[19] Brummer N., de Villiers E. 'The BOSARIS toolkit user guide: theory, algorithms and code for binary classifier score processing'. 2011;i:1–24.

[20] Alam J., Boulianne G., Burget L. 'Analysis of ABC submission to NIST SRE 2019 CMN and vast challenge'. *Proc.* Odyssey 2020 The Speaker and Language Recognition Workshop; 2020. pp. 289–95.

[21] Landini F., Wang S., Diez M., *et al.* 'But system description for DIHARD speech diarization challenge 2019'. *arXiv preprint arXiv:191008847*. 2019.

[22] Petrovska-Delacrétaz D., Chollet G., Dorizzi B. *Guide to Biometric Reference Systems and Performance Evaluation.* Springer; 2009.

[23] Huang G.B., Learned-Miller E. *'Labeled faces in the wild: Updates and new reporting procedures'*. Dept Comput Sci, Univ Massachusetts Amherst, Amherst, MA, USA, Tech Rep; 2014. pp. 3–14.

[24] Learned-Miller E., Huang G.B., Roy CA., Li H., Hua G. LFW : results [online]. 2019. Available from http://vis-www.cs.umass.edu/lfw/results.html [Accessed 12 May 2021].

[25] He K., Zhang X., Ren S., *et al.* 'Deep residual learning for image recognition'. Proceedings of the IEEE Conference on Computer Vision and Pattern Recognition; 2016. pp. 770–8.

[26] Schroff F., Kalenichenko D., Philbin J. 'Facenet: a unified embedding for face recognition and clustering'. Proceedings of the IEEE Conference on Computer Vision and Pattern Recognition; 2015. pp. 815–23.

[27] Taigman Y., Yang M., Ranzato M., *et al.* 'Deepface: closing the gap to human-level performance in face verification'. Proceedings of the IEEE Conference on Computer Vision and Pattern Recognition; 2014. pp. 1701–8.

[28] Sun Y., Liang D., Wang X., *et al.* 'Deepid3: face recognition with very deep neural networks'. *arXiv preprint arXiv:150200873*. 2015.

[29] Parkhi O.M., Vedaldi A., Zisserman A. 'Deep face recognition'. BMVC: proceedings of the British Machine Vision Conference; 2015. p. 1.

[30]   Yi D., Lei Z., Liao S., *et al.* 'Learning face representation from scratch'. *arXiv preprint arXiv:14117923*. 2014.

[31]   Amos B., Ludwiczuk B., Satyanarayanan M. 'Openface: a general-purpose face recognition library with mobile applications'. *CMU School of Computer Science*. 2016;6.

[32]   Deng J., Guo J., Zhou Y., *et al. RetinaFace: single-stage dense face localisation in the wild*. 2019. Available from http://arxiv.org/abs/1905.00641 [Accessed 12 May 2021].

[33]   King D.E. 'Dlib-ml: a machine learning toolkit'. *Journal of Machine Learning Research*. 2009;10:1755–8.

[34]   Viola P., Jones M. 'Rapid object detection using a boosted cascade of simple features'. Proceedings of the 2001 IEEE Computer Society Conference on Computer Vision and Pattern Recognition. CVPR 2001. vol. 1. IEEE; 2001. p. I.

[35]   Yang S., Luo P., Loy C.C., *et al.* 'WIDER FACE: a face detection benchmark'. IEEE Conference on Computer Vision and Pattern Recognition (CVPR); 2016.

[36]   Moschoglou S., Papaioannou A., Sagonas C. 'AgeDB: the first manually collected, in-the-wild age database'. IEEE Computer Society Conference on Computer Vision and Pattern Recognition Workshops; 2017.

[37]   Guo Y., Zhang L., Hu Y., *et al.* 'MS-celeb-1M: A dataset and benchmark for large-scale face recognition'. European Conference on Computer Vision; 2016. pp. 87–102.

[38]   Ester M., Kriegel H.P., Sander J., *et al.* 'A density-based algorithm for discovering clusters in large spatial databases with noise'. Kdd; 1996. pp. 226–31.

[39]   Hmani M.A., Petrovska-Delacrétaz D. 'State-of-the-art face recognition performance using publicly available software and datasets'. 2018 4th International Conference on Advanced Technologies for Signal and Image Processing (ATSIP). IEEE; 2018. pp. 1–6.

[40]   Gross R., Matthews I., Cohn J., Kanade T., Baker S. 'Multi-pie'. *Image and Vision Computing*. 2010;28(5):807–13.

[41]   Li B., Xiong W., Hu W., *et al.* 'SSD: Single shot multibox detector'. *European Conference on Computer Vision. Springer*; 2016. pp. 21–47.

[42]   Kazemi V., Sullivan J. 'One millisecond face alignment with an ensemble of regression trees'. Proceedings of the IEEE Computer Society Conference on Computer Vision and Pattern Recognition; 2014. pp. 1867–74.

[43]   Bulat A., Tzimiropoulos G. 'How far are we from solving the 2D & 3D face alignment problem? (and a dataset of 230,000 3D facial landmarks)'. *Proceedings of the IEEE International Conference on Computer Vision*. 2017;2017:1021–30.

[44]   Sagonas C., Antonakos E., Tzimiropoulos G., Zafeiriou S., Pantic M., *et al.* '300 faces in-the-wild challenge: database and results'. *Image and Vision Computing*. 2016;47(6):3–18.

[45]   Szegedy C., Vanhoucke V., Ioffe S., *et al.* 'Rethinking the inception architecture for computer vision'. Proceedings of the IEEE conference on computer vision and pattern recognition; 2016. pp. 2818–26.

[46] Szegedy C., Ioffe S., Vanhoucke V., *et al.* 'Inception-v4, inception-ResNet and the impact of residual connections on learning'. 31st AAAI Conference on Artificial Intelligence, AAAI 2017; 2017. pp. 4278–84.

[47] Lin C.J., Weng R.C., Keerthi S.S. 'Trust region newton methods for large-scale logistic regression'. Proceedings of the 24th international conference on Machine learning; 2007. pp. 561–8.

[48] Marras M., Marín-Reyes P.A., Lorenzo-Navarro J., *et al.* 'Deep multi-biometric fusion for audio-visual user re-identification and verification'. *International Conference on Pattern Recognition Applications and Methods. Springer*; 2019. pp. 136–57.

Chapter 10

Voice biometrics: future trends and challenges ahead

Douglas Reynolds[1] and Craig S. Greenberg[2]

Voice has become woven into the fabric of everyday human–computer interactions via ubiquitous assistants like Siri, Alexa, Google, Bixby, Viv, etc. The use of voice will only accelerate as speech interfaces move to wearables [1], vehicles [2], and IoT devices and appliances [3]. As a person's voice is used more to control real-world actions and access private information, the role of voice biometrics will play an increasingly important role in protecting sensitive access and actions [4] and providing personalization for services and devices [5]. This widespread use will bring many new application opportunities as well as challenges in addressing societal privacy concerns [6], securing systems against sophisticated attacks [7], and continuously improving voice biometric's reliability in more diverse acoustic environments [8]. In this chapter we will present our assessment of future trends in these important areas.

10.1 Applications

There are many application areas that are currently or soon will be taking advantage of voice biometrics. While a majority of current applications are focused on banking access [9], call center fraud prevention [10], and multiuser assistant personalization [11], there are growing uses in the fields of forensics [12], boarder control [13], government benefits registration [14], and healthcare [15]. Below, we describe two trends we expect to see coming to many voice biometric applications.

Continuous trust monitoring: A majority of voice biometric authentication applications today use a discrete step where a person is asked to provide a voice sample, which is compared to a model of the proffered identity and the match score is compared to a threshold to decide whether to allow or deny access. This is often a *text-dependent* system that uses a predefined or prompted phrase [16]. With this

[1]MIT Lincoln Laboratory, Massachusetts, United States
[2]NIST, Maryland, United States

explicit authentication step, once a person passes, he/she is allowed access for all account actions regardless of the sensitivity.[1]

When a system is using voice navigation, there are multiple speech samples collected that can be used to provide a continuously updated match score. In this continuous trust monitoring system [17, 18], not only is the authentication made implicit and seamless to the user, but also it is possible to re-authenticate with a more stringent threshold when more sensitive actions are requested. Such a system may also be useful when a human operator is working with a customer to provide a voice trust score that can prompt the operator to seek more confirmatory information before providing further sensitive information.

For shared home devices or vehicles, the continuous trust monitoring can be used to detect when a different user is accessing the device at a particular time [19]. The trust score can then be used to personalize the accessed service (e.g., playlist, search history, etc.) and authenticate purchase authority, perhaps having higher purchase amounts requiring higher trust.

Multimodal authentication: With cameras and increased compute power, the default on newer devices (even those without real estate for screens), there will be a trend to including face authentication in conjunction with voice authentication. Face authentication is on its own a strong biometric with a great deal of supporting research and highly accurate algorithms [20]. However, with people using wearable and mobile devices in less-controlled and more harsh real-world environments, strong authentication will rely on multiple independent biometric data streams to maintain high accuracy [21]. The promise here is not only that fusion of face and voice provides improved overall accuracy, as described in this chapter, but also that the system can back off to the best modality under degraded conditions (poor audio/ good image, good audio/poor image) [22].

Other device sensor data could also provide independent biometric data streams for fusion, such as gate estimation [23], pattern of life [24], app usage [25], typing speed/style [26], etc. However, there remain many open privacy and ethical considerations when contemplating using sensor fingerprinting for identification purposes [27].

10.2   Privacy and security

As voice biometrics become more widespread, privacy and security concerns will need to be clearly addressed. Chapters 5 and 6 in this book described aspects of privacy in the health-care domain and the need for de-identification algorithms when anonymity is required. Other aspects of privacy such as governance of when, where, and how voice biometrics are allowed and the use of a person's voice biometric by

---

[1]Of course, multifactor one-time codes can be used to verify for more sensitive activities, but this too requires another explicit user step.

government and third parties should be part of the larger discussion of any biometric's use in society. Voice, unlike face and fingerprint, carries with it the additional privacy concern of protecting the content of the spoken message in addition to just who is associated with speaking it.

In terms of security, as voice biometrics proliferate and guard access to more valuable information, they naturally will become the focus of more hacking attempts. Similar to cybersecurity, securing voice biometrics will be an ongoing, escalating cat-and-mouse game of sophisticated attacks and comprehensive defenses. Unlike keeping passwords private, our voice is openly available, making voice biometrics vulnerable to some simple attacks, such as replay [28]. Luckily, there are layered security approaches (e.g., prompted random phrases) and algorithmic approaches (e.g., Chapter 4) to mitigate this attack. However, the growing areas of personal voice synthesis and adversarial machine learning pose considerable threats to voice biometrics in the future.[2]

In the security cat-and-mouse game, the cost and the specialized technical knowledge necessary to pull off an effective attack can often be a mitigating factor in favor of the defender. As the barrier to entry of an attack decreases, the threat of it being successfully used increases. At this point in the research world of voice synthesis and adversarial machine learning, not only are better techniques rapidly being developed, code, models, and data are also widely and freely available, dramatically decreasing the barrier to entry for would-be attackers. This genie cannot be put back in the bottle, but similar conditions means research efforts focused on addressing these vulnerabilities are becoming equally available for defenders to use. Going forward, deployed voice biometrics will need to be equipped to defend not only known attacks but also have the ability to rapidly adapt new defenses. These defenses will come in the form of detecting a suspected synthetic or attack sample and/or building the voice biometric to be robust to such attacks, potentially incorporating synthetic and attack sample into the training data [29]. One implication of this need will be that voice biometric developers may need to employ red-teams who are skilled at finding and implementing credible attacks against their systems to better harden their deployed systems. Another potential defense is to incorporate more speaker-specific behavioral and content features, such as idiolect and prosody, that may be more difficult to spoof [30].

10.3 Research

The accuracy of voice biometrics has improved dramatically over the past decade, driven by the use of more advanced machine learning algorithms and large

---

[2]https://www.blog.google/outreach-initiatives/google-news-initiative/advancing-research-fake-audio-detection

amounts of speaker annotated data to train these recognizers. And while today's accuracy is sufficient for many current applications, as voice biometrics move to wearables, vehicles, and IoT devices, they will need to become far more robust in these harsh audio environments [31]. However, the known approach of collecting data from a new environment to retrain and tune systems is cumbersome and does not scale to the growing number of diverse acoustic environments encountered. Further, as discussed in Section 10.2, they will need to cope with synthetic and adversarial attacks. Systems will need to be more adaptive and field-trainable to meet these challenges. Beyond these, we expect to see recent research trends in machine learning and other biometrics to influence the directions of voice biometric research. Here are a few general research areas we see gaining importance going forward.

Learning with limited labels: There is an extremely large (and growing) amount of audio data available covering a wide variety of audio environments. The catch is that it is not labeled by speaker, thus making it difficult or impossible to use for training and testing voice biometric algorithms. Research into how to successfully use unlabeled or limited labels [32], possibly with errorful labels, will potentially allow algorithms to tap into this vast and varied data to build more accurate and robust voice biometric systems.

Monitoring speaker model mismatch and health: As voice biometrics are being used in more varied audio environments, the potential for mismatch between the audio environments used to enroll the speaker's model and that in which it is being tested increase. This mismatch of course can cause more speakers to be falsely rejected but it can also cause spurious false accepts, both of these sometimes with very high confidence. An automatic means to detect a mismatch would allow for ways to alert the user of a potential problem or trigger mitigation strategies such as promoting other models of authentication. A related area of research is automatic means to measure the health of speaker model. Speaker models may be updated continuously and sometimes implicitly using conversation interactions with an assistant [33]. Knowing that a speaker model is drifting or has been updated with poor data would allow for prompting for a fresh enrollment to maintain accuracy for the user.

Achieving audio environment independence: Current voice biometric systems can be rather brittle in that they work extremely well when used in an audio environment similar to that on which they were trained, but fail quickly as the test audio environment diverges. A long-time, but elusive, research goal is to develop voice biometric algorithms that are agnostic to the audio environment and focused exclusively on the speaker characteristics. This is of course very difficult because the acoustics of the environment and the speaker are intimately intertwined at the signal level. Some of the above research areas may prove useful here, as would techniques like audio environment transfer learning (learn to transfer the audio characteristics from one environment to another, analogous to style transfer learning for images) and synthetic generation of training data driven by some environment parameters [34].

Addressing the attack threat: As discussed in this book, there are looming threats to voice biometrics coming from ever improving personal voice synthesis and advanced adversarial machine learning. The two main ways to defend against these threats is to detect when they are occurring to take appropriate action or to build in some level of immunity to the attacks into the voice biometric algorithm. While detection has the advantage of being a separate system from the biometric that can be updated rapidly, it relies on finding some flaws or signatures in the attack audio that can be used to detect their presence, which can be fleeting at the pace attacks change. Making systems robust to the attack has the potential of a longer lasting defense but is more difficult to enact. This requires a broad view of vulnerabilities in the voice biometric algorithm, not just for particular attacks.

Explainability: There is an ongoing and recently trending area in machine learning research focusing on providing explanations for machine learning system output ([35] both analyzes the growth in research interest in explainable AI and has itself been cited more than 270 times in the last six months). There are many types of applications where providing an explanation of the system output is useful, for example providing explanations to regulators in regulatory environments, to downstream users when they are affected by the model, or to model developers as a tool to better understand the model itself (see Table 1 in [35]). While voice biometrics might be utilized in the context of each of these types of applications, explainability seems particularly suited for forensic applications, where there is often a need to be able to explain how a determination is reached.

Bias: The differences in performance of deployed facial recognition systems based on the demographics of the subjects being recognized have been discovered and highlighted as problematic [36]. Due to the similarities between face biometrics and voice biometrics, it seems reasonable to conjecture that many of these same issues might be encountered in cases where voice biometrics are similarly deployed. In such a case, appropriately dealing with bias in voice biometrics will be an increasingly important research area, particularly as voice biometric systems are deployed in new and growing applications.

Prognostication of technology trends is at best an educated guess, and in this chapter we have provided our best guess on trends in voice biometrics. While the details may not play out as predicted, the broad directions in growing, novel applications, increasing scrutiny on privacy and security, and wider research efforts into robust systems will most certainly materialize.

References

[1] Starner T.E. 'The role of speech input in wearable computing'. *IEEE Pervasive Computing.* 2002;1(3):89–93.

[2] Barón A., Green P. *Safety and usability of speech interfaces for in-vehicle tasks while driving: A brief literature review*. Technical Report UMTRI-2006-5. University of Michigan Transportation Research Institute; 2006.

[3] Mehrabani M., Bangalore S., Stern B. 'Personalized speech recognition for Internet of Things'. *2015 IEEE 2nd World Forum on Internet of Things WF-IoT*. IEEE; 2015. pp. 369–74.

[4] Morgen B. 'Voice biometrics for customer authentication'. *Biometric Technology Today*. 2012;2012(2):8–11.

[5] Ghosh S., Pherwani J. 'Designing of a natural voice assistants for mobile through user centered design approach'. *International Conference on Human-Computer Interaction*. Springer; 2015. pp. 320–31.

[6] Pal D., Arpnikanondt C., Razzaque M.A. 'Personal information disclosure via voice assistants: the personalization–privacy paradox'. *SN Computer Science*. 2020;1(5):1–17.

[7] Sanchez-Casanova J., Goicoechea-Telleria I., Liu-Jimenez J., *et al.* 'Performing a presentation attack detection on voice biometrics'. *2018 International Carnahan Conference on Security Technology ICCST*. IEEE; 2018. pp. 1–5.

[8] Scheffer N., Ferrer L., Lawson A., *et al.* 'Recent developments in voice biometrics: robustness and high accuracy'. *2013 IEEE International Conference on Technologies for Homeland Security HST*. IEEE; 2013. pp. 447–52.

[9] Aronowitz H., Hoory R., Pelecanos J., *et al.* 'New developments in voice biometrics for user authentication'. Twelfth Annual Conference of the International Speech Communication Association; 2011.

[10] Markowitz J.A. 'Voice biometrics'. *Communications of the ACM*. 2000;43(9):66–73.

[11] Nurgaliyev K., Di Mauro D., Khan N., *et al.* 'Improved multi-user interaction in a smart environment through a preference-based conflict resolution virtual assistant'. *2017 International Conference on Intelligent Environments IE*. IEEE; 2017. pp. 100–7.

[12] Dellwo V., French P., He L. 'Voice biometrics for FORENSIC speaker recognition applications' in Frühholz S., Belin P. (eds.). *The Oxford Handbook of Voice Perception*. Oxford: Oxford University Press; 2018. pp. 777–98.

[13] Sequeira A.F., Chen L., Ferryman J., *et al.* 'PROTECT multimodal DB: fusion evaluation on a novel multimodal biometrics dataset envisaging border control'. *2018 International Conference of the Biometrics Special Interest Group (BIOSIG*. IEEE; 2018. pp. 1–5.

[14] Vallie Z. New biometric verification for SASSA grant recipients. Independent Online (South Africa)n. 2021. Available from https://www.iol.co.za/business-report/economy/new-biometric-verification-for-sassa-grant-recipients-15102561 [Accessed 10 May 2021].

[15] Sigona F. 'Voice biometrics technologies and applications for healthcare: an overview'. *JDREAM Journal of InterDisciplinary Research Applied to Medicine*. 2018;2(1):5–16.

[16] Variani E., Lei X., McDermott E., *et al.* 'Deep neural networks for small footprint text-dependent speaker verification'. *2014 IEEE International Conference on Acoustics, Speech and Signal Processing ICASSP*. IEEE; 2014. pp. 4052–6.

[17] Campbell J., Campbell W., Jones D. 'Biometrically enhanced software-defined radios'. *in 2003 Software Defined Radio Technical Conference.* Wireless Innovations Forum; 2003. pp. 447–52.

[18] Hazen T., Jones D., Park A., *et al.* 'Integration of SPEAKER recognition into conversational spoken dialogue systems'. *Eurospeech. ISCA.* 2003:1961–4.

[19] Feng H., Fawaz K., Shin K.G. 'Continuous authentication for voice assistants'. *Proceedings of the 23rd Annual International Conference on Mobile Computing and Networking*; 2017. pp. 343–55.

[20] Masi I., Wu Y., Hassner T., *et al.* 'Deep face recognition: a survey'. *2018 31st SIBGRAPI conference on graphics, patterns and images (SIBGRAPI).* IEEE; 2018. pp. 471–8.

[21] Blasco J., Chen T.M., Tapiador J., *et al.* 'A survey of wearable biometric recognition systems'. *ACM Computing Surveys.* 2016;49(3):1–35.

[22] Omid Sadjad S., Greenberg C., Singer E. *The 2019 NIST audio-visual SPEAKER recognition evaluation.* Odyssey: ISCA; 2020.

[23] Wang L., Ning H., Tan T., *et al.* 'Fusion of static and dynamic body biometrics for gait recognition'. *IEEE Transactions on Circuits and Systems for Video Technology.* 2004;14(2):149–58.

[24] Yuan M., Nara A. 'Space-time analytics of tracks for the understanding of patterns of life'. *in Space-time Integration in Geography and GIScience.* Springer; 2015. pp. 373–98.

[25] Bo C., Zhang L., Jung T., *et al.* 'Continuous user identification via touch and movement behavioral biometrics'. *2014 IEEE 33rd International Performance Computing and Communications Conference IPCCC.* IEEE; 2014. pp. 1–8.

[26] Flior E., Kowalski K. 'Continuous biometric user authentication in online examinations'. *2010 Seventh International Conference on Information Technology: New Generations.* IEEE; 2010. pp. 488–92.

[27] Mordini E., Tzovaras D. *Second Generation Biometrics: The Ethical, Legal and Social Context.* 11. Springer Science & Business Media; 2012.

[28] Lavrentyeva G., Novoselov S., Malykh E., *et al.* 'Audio replay attack detection with deep learning frameworks'. *Interspeech*; 2017. pp. 82–6.

[29] Saratxaga I., Sanchez J., Wu Z., *et al.* 'Synthetic speech detection using phase information'. *Speech Communication.* 2016;81:30–41.

[30] Reynolds D., Andrews W., Campbell J., *et al.* 'The SuperSID project: exploiting high-level information for high-accuracy speaker recognition'. *International Conference on Acoustics, Speech, and Signal Processing.* IEEE; 2003.

[31] Qin X., Li M., Bu H. 'The INTERSPEECH 2020 far-field SPEAKER verification challenge'. *Interspeech. ISCA.* 2020.

[32] Huh J., Heo H., Kang J. 'Augmentation adversarial training for unsu-pervised SPEAKER recognition'. *in Workshop on Self-supervised Learning for Speech and Audio Processing.* NeurIPS; 2020.

[33] Barras C C., Meignier S., Gauvain J. *Unsupervised Online Adaptation for SPEAKER Verification over the Telephone. Odyssey*: ISCA; 2004.

[34]   Yi Liu Y., Zhuang B., Li Z. 'Cross-domain SPEAKER recognition us-
       ing cycle-consistent adversarial networks'. Asia-Pacific Signal and
       InformationProcessing Association Annual Summit and Conference. IEEE;
       2019.

[35]   Barredo Arrieta A., Díaz-Rodríguez N., Del Ser J., *et al*. 'Explainable artifi-
       cial intelligence (XaI): concepts, taxonomies, opportunities and challenges
       toward responsible AI'. *Information Fusion*. 2020;58(3):82–115.

[36]   Nature. Is facial recognition too biased to be let loose [online]? Available
       from https://www.nature.com/articles/d41586-020-03186-4 [Accessed 16
       Dec 2020].

Index